SPEED READ

PORSCHE 911

Quarto.com

© 2018 Quarto Publishing Group USA Inc. Text © 2018 Wayne R. Dempsey

First Published in 2018 by Motorbooks, an imprint of The Quarto Group, 100 Cummings Center, Suite 265-D, Beverly, MA 01915, USA. T (978) 282-9590 F (978) 283-2742

Motorbooks titles are also available at discount for retail, wholesale, promotional, and bulk purchase. For details, contact the Special Sales Manager by email at specialsales@quarto.com or by mail at The Quarto Group, Attn: Special Sales Manager, 100 Cummings Center, Suite 265-D, Beverly, MA 01915, USA.

ISBN: 978-0-7603-6322-5

Digital edition published in 2018
eISBN: 978-0-7603-6323-2

Library of Congress Cataloging-in-Publication Data is on file.

Acquiring Editor: Darwin Holmstrom
Project Manager: Alyssa Lochner
Art Director: James Kegley
Illustrations: Jeremy Kramer

SPEED READ

PORSCHE 911

THE HISTORY, TECHNOLOGY AND DESIGN
BEHIND GERMANY'S LEGENDARY SPORTS CAR

WAYNE R. DEMPSEY

motorbooks

INTRODUCTION

The Porsche 911 is one of the world's most recognizable car models and a cultural icon. Like the Nike swoosh, or the red swirl of Coca Cola, the shape of the 911 is instantly recognizable to enthusiasts worldwide. But the 911 is more than just an iconic design; it has also backed up its reputation with decades of winning performances on the world's greatest racetracks. Its precision German engineering, finely tuned performance, and styling-with-purpose aesthetic all combine to make the Porsche 911 one of the world's most popular sports cars. Indeed, the ownership of a 911 tops many a bucket list.

Speed Read: Porsche 911 is designed as a reference for beginners and experts alike. For beginners new to the wonderful and unique world of the Porsche 911, this book covers the basics—the who, what, when, where, and why of the 911 and how it came to be the legend that it is today. For experts, this book offers the opportunity to delve into the stories of the rainmakers, the tales of the cars, the influencers, and the heroes at Porsche whose stories aren't always covered in 911 books. Finally, the book's illustrations are simply awesome, the perfect companion to the text.

Though organized around themes, Speed Read books invite exploration and do not need to be read front to back. The early chapters of this book cover the evolution of the 911 from its initial design to its current form. The heart of the book details the Porsche 911's racing success combined with tales of some of the greatest 911s ever produced. The final chapter looks at the passionate community that has evolved around the 911 from its earliest days.

Who am I? I founded an Internet-based company called Pelican Parts about twenty years ago, and our main goal is to supply enthusiasts with parts and knowledge so they can repair their Porsches. I've written a handful of technical books on the 911, essential reading material when you finally do take the plunge and pick up one of these wonderful cars. And, like you, I love the 911.

THE PEOPLE'S SPORTS CAR

THE PEOPLE'S SPORTS CAR
FERDINAND AND FERRY PORSCHE

FUN FACT

Ferdinand Porsche is credited with creating the world's first hybrid electric car—in 1899! The Lohner-Porsche "Semper Vivus" was essentially an electric car with an internal combustion engine combined with a generator. The car had all-wheel drive and four-wheel brakes—the first of its kind.

HISTORICAL TIDBIT

Porsche officials inked an important contract in the spring of 1949. As part of an effort by Volkswagen to improve the design of the Beetle, Porsche earned a royalty on every car produced, and gained access to Volkswagen's network of dealers and service centers. Volkswagen also agreed to supply the raw materials for 356 manufacturing. This became a huge windfall for Porsche as more than 20 million Type I Beetles were built.

KEY PERSON

Aton Piëch was the son-in-law of Ferdinand Porsche. The Piëch and Porsche families feuded for decades.

It all started with Ferdinand, the founder of Porsche and patriarch of the famous family. Ferdinand Porsche founded the Porsche design and consulting firm in 1931. In 1934, under the direction of the German government, he designed the original Volkswagen Beetle. Known as the "people's car" (translated into the term "Volkswagen"), its design was similar to his previous work on the 1931 Zündapp Type 12 car and the Tatra V570, both of which featured a rear-engine design with air-cooled engines.

After World War II, Ferdinand Porsche had difficulty restarting his company in the wake of post-war embargoes and reparations. To assist, Ferry Porsche, Ferdinand's son, took the helm of the company with his sister Louise. In 1948, Porsche built its first car, the original 356, which was a two-seater, open-top roadster with a mid-engine layout. At the same time, they developed the 356/2—a unibody, rear-engine design that closely echoed the design of the Volkswagen Beetle.

Porsche marketed the new 356 sports car to Volkswagen dealers and it proved popular with wealthy customers. About fifty early, handcrafted aluminum cars were built inside a rickety old sawmill in the town of Gmünd. More than 77,000 units were sold over seventeen years of production, establishing the foundation for Porsche's future success.

Ferry Porsche would also kickstart the company's racing efforts. On July 11, 1948, a Porsche 356 made history with a class victory in the Austrian Rund um den Hofgarten, a rally through the residential streets of the ski town of Innsbruck.

In 1972, amid internal family feuding, Ferry Porsche transformed the company from a limited partnership into a public company (see "Separating Family from Company," page 20).

ADDITIONAL READING: *Cars Are My Life*, by Ferry Porsche

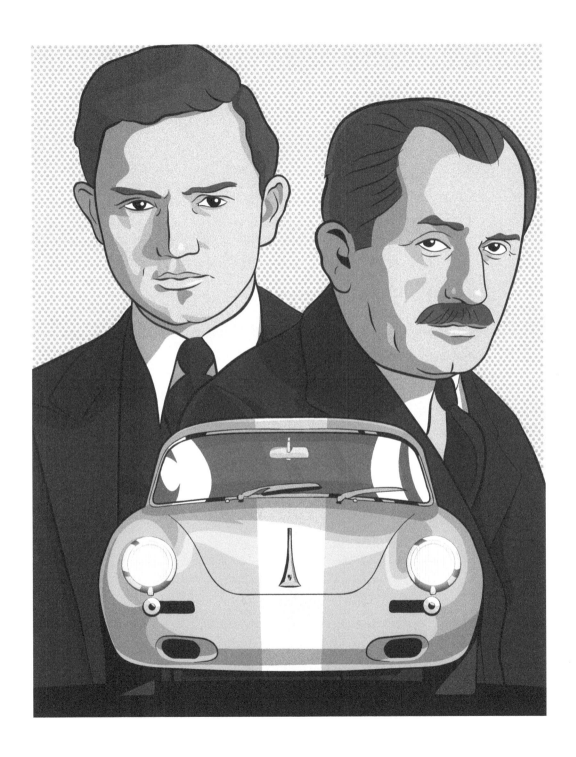

THE PEOPLE'S SPORTS CAR
PORSCHE 356

The Porsche 356 was based upon founder Ferdinand "Ferry" Porsche's 1948 design of a hand-built, mid-engine cabriolet, and it established the company's reputation for engineering prowess and racing success. The 356 became a German icon and put Porsche's name on the map.

Ferry Porsche was the principle designer of the Volkswagen Beetle, and the similarities between the 356 and the Volkswagen "bug" are numerous. The two cars share their small size, air-cooled rear engine design, two-door configuration, and numerous parts. The Porsche 356 engine is a higher-performance version of the air-cooled Volkswagen Type 1 engine, complete with similar push rods, upright fan cooling, center-mounted camshaft, and big, air-cooled cylinders.

The shape of the 356 has evolved slightly over the years, but its signature design element— the sloping rear deck over the engine bay—endured and foreshadowed the look and feel of the Porsche 911 in future years.

The 356 was lightweight, nimble, well designed, and well constructed. Although very expensive at the time, the 356 earned a large following; Porsche ultimately sold more than 77,000 units from 1948 to 1965. As with the early Porsche 911, the 356 has a unique cachet with owners who appreciate the German engineering, handling, and overall uniqueness of the car.

The Porsche 901 (soon to be renamed the 911) made its debut at the 1963 Frankfurt Motor Show and was a natural evolution of the 356 design; Porsche needed to alter the chassis and expand the engine compartment to make room for the more powerful six-cylinder "Mezger" engine developed for the 911.

ADDITIONAL READING: *Porsche Speedster Typ 540: Quintessential Sports Car,* by Steve Heinrichs, Marco Marinello, Jim Perrin, Lee Raskin, Charles A. Stoddard, and Don Zingg

THE PEOPLE'S SPORTS CAR
BUTZI PORSCHE

FUN FACT
The fifty-seven-story-tall Porsche Design Tower that opened in Sunny Isles, Florida, in 2017 features a robotic parking garage elevator that transports residents' cars right to their apartments.

HISTORICAL TIDBIT
The first Porsche Design product was the Chronograph 1, the world's first all-black wristwatch. It was designed to resemble the dashboard gauges of the 1974 Porsche 911.

KEY PERSON
Wendelin Wiedeking is the Porsche CEO who oversaw the company's return from a protracted decline. Butzi Porsche was chairman of the supervisory board at the time Wiedeking was hired.

I'm guessing when they flunked Ferdinand Alexander Porsche out of design school, they had no idea he would go on to design the world's most recognizable sports car. After leaving school, Ferdinand, best known as F. A. or Butzi, joined the family business and went on to design the Porsche 904 and Porsche 911.

Butzi got his start at the Porsche workshop, spending Sundays watching his father and grandfather solve engineering issues. After a hiatus from Germany during World War II, Butzi returned to attend the Ulm School of Design, but didn't make it past the second semester. He joined the family's company in 1957, just as redesign work was starting on the flagship 356; this design would ultimately become the 911. Butzi is credited with designing and sculpting a Plasticine model that was well received by his father and others in the design department. But politics intervened. Erwin Komenda, the head of Porsche's body design department, preferred the designs of Graf Goertz (designer of the BMW 507), and didn't cooperate with the implementation of young Butzi's design. The model was built across the street by coachbuilder Reutter instead.

Ousted by Ferry Porsche in a 1972 family shake-up, Butzi started his own firm, Porsche Design Studio, in Zell, Austria. The new company designed industrial and consumer products including sunglasses, luggage, watches, and electronics. Porsche Design proved to be immensely successful, and it merged with Porsche AG in 2003, but remained an independent division and performed design work for other clients.

Butzi returned to Porsche AG as chairman of the supervisory board in 1990, a time when the company was suffering from slow sales and stale models. He spearheaded the hiring of talented new management that oversaw the introduction of the Boxster and the modernization of the company's assembly lines.

THE PEOPLE'S SPORTS CAR
EARLY 911 LONG HOODS (1964–1973)

FUN FACT

The ignition switch is on the left side of the steering wheel in Porsche vehicles, a throwback to racing the 24 Hours of Le Mans. In the race's early years, drivers had to run to their cars, start them, and pull away. Porsche learned that placing the ignition switch on the left-hand side let drivers start and shift at the same time, saving precious seconds at the start.

HISTORICAL TIDBIT

The first cars weren't called the 911, they were called the 901. However, Porsche learned that in many key markets, Peugeot held the trademark on car names with a zero as the middle digit. Porsche changed the name of the car to 911 before the cars were sold to the public.

KEY PERSON

Ferdinand Alexander Porsche was credited with the original Porsche 911 design.

The first-generation 911 models hold a special place in many owners' hearts. These early models are typically referred to as "long hoods," mainly because the hood was shortened in 1974 due to new bumper regulations.

When the first 911 was released to the public in 1964, early 2.0-liter engines were considered very robust. But with 130 horsepower, they were not considered terribly quick. In 1967, Porsche released the first performance version, the 1967 911 S, which sported Fuchs alloy wheels for the first time and delivered about 25 percent more horsepower than the base model 911.

For 1968, three different performance levels were available: the 911 T with a detuned version of the engine), the 911 L (similar to the previous year's base engine), and the 911 S (top of the line in horsepower). These three strata of 911 performance were available until 1973 (with the 911 L being renamed the 911 E in 1969). All three were nearly identical except for trim, engine options, and fuel injection systems, and halfway through 1973, Porsche switched the fuel injection on the 911 T to CIS (Continuous Injection System), creating a half-year oddball model.

These early 911s were once considered underpowered, rust-prone, and generally primitive compared to later models such as the 911SC and Carrera. However, their collectability has in recent years increased, catapulting the long hood's desirability to the top of the 911 food chain. These early cars are now the most valuable and most sought-after, with upscale auction houses often selling well-preserved examples for six-figure sums.

One legend in the 911 world involves the 911 ST. Although the factory never officially labeled any 911 as a 911 ST, the designation generally refers to specially ordered cars that were ordered with a 911 S drivetrain, and 911 T (lightweight) trim. These 911 ST cars were often spec'd out by privateer racing teams.

THE PEOPLE'S SPORTS CAR
TARGA

When launched in 1965, the Porsche 911 was only available as a coupe. With the previous 356 model, Porsche had sold many open-top cars (called cabriolets), and the public clamored for a 911 equivalent. However, the mid-1960s were a time of uncertainty for the automotive industry. Ralph Nader's book *Unsafe at Any Speed* had called attention to shortcomings in car safety, and convertibles, which lacked rollover protection, were targeted as potentially unsafe.

With convertible design legislation pending, Porsche set out to develop an alternative. Thus, in 1967, the 911 Targa was born. Porsche named this innovative body style after the famous race on the island of Sicily, the Targa Florio. A large roll bar was located just rearward of the doors, replacing the typical B-pillars of a coupe. The rear window of early models was made of canvas and clear plastic, and these cars were called "soft window Targas." A year later in 1968, Porsche ditched the canvas and replaced it with a solid glass rear window.

The fixed-window Targa design was produced until 1994, when the new 993 was introduced. In 1996, the Porsche 911 Targa (993) featured a unique redesign that replaced the distinctive roll bar with a large, panoramic glass roof that slid backward to effectively create a large sunroof. This design continued with the 996 and 997, with Porsche adding the ability to open the glass like a hatchback. Finally, in 2014, Porsche returned to its roots, creating a retro-look Targa that featured the classic Targa bar design. This model incorporated a complex mechanism that stowed the roof under the fixed rear glass window and engine lid.

Targa fans are passionate about their cars and love the open-air motoring without the drawbacks of extremely windy conditions.

THE PEOPLE'S SPORTS CAR
SEPARATING FAMILY FROM COMPANY

Up until the early 1970s, Porsche was a family-run business. But in 1972, amid family conflict, Dr. Ferry Porsche made changes to convert the company into a partly shareholder-owned corporation, Porsche AG.

This massive change meant company management would be chosen solely based on performance, not family heritage, and there were some downsides. At least some of the Porsche family members were quite good at what they did, and the changes at Porsche caused them to leave. For example, F. A. "Butzi" Porsche left and founded his own company, Porsche Design Studio.

Ferdinand Piëch, who had been instrumental in developing Porsche's engineering department, also left to join Audi-NSU, where he would excel and end up leading the entire VW-Audi conglomerate. Replacing Piëch was Ernst Fuhrmann, who firmly believed Porsche's racing cars should be derived from current production cars, which was good news for the 911.

The events of 1972 set the stage for an epic showdown in 2008 between Ferdinand Piëch and Wendelin Wiedeking, the CEO of Porsche. In a bold, aggressive move, starting in mid-2005, Porsche AG began acquiring options to purchase shares of Volkswagen. Wiedeking used Porsche's credit to borrow heavily to purchase the shares and options. In 2008, following an economic downturn, Porsche faced a liquidity crisis when banks refused to lend it any more money. Wiedeking turned to the emir of Qatar for assistance, but the deal was scuttled by Piëch, who convinced the emir to invest directly in Volkswagen instead. The ironic result of this battle was a cash-starved, debt-laden Porsche AG, which was forced to agree in 2009 to a corporate takeover by Volkswagen in one of the decade's most bizarre financial stories.

THE PEOPLE'S SPORTS CAR
G-SERIES (1974–1989)

The mid-1970s were difficult years for the Porsche 911 in America. Impact bumpers added heft and weight, and emissions equipment curtailed engine life. The 1974–1977 engines had a magnesium block that was prone to overheating and oil leaks, and required expensive repairs. In 1975–1976, Porsche set the stage in Europe for a huge upgrade with both the Euro Carrera 3.0 and the 911 Turbo. With a few minor changes, the Euro Carrera was rebranded for the US market in 1978 as the 911SC.

The 911SC is a great 911 model. By the time of its debut, 911 design was well refined and the cars were reliable, sporty, and just plain fun. With its CIS-injection and warm-up circuits, it was the first Porsche 911 you could simply hop in and turn the key to start it up—no pumping of the gas pedal or pulling hand throttles required. The 911SC bodies were galvanized so they didn't rust as much as earlier 911s. The well-developed 3.0-liter engine in the 911SC could run for hundreds of thousands of miles between rebuilds—if a few minor reliability updates were made (namely, improved chain tensioners, and an air box pop-off valve). About 58,000 911SCs were produced, with about 25,000 sent to North America, so 911SC models remain relatively plentiful and reasonably priced compared to earlier 911s. I recommend a well-sorted 911SC to a prospective 911 buyer all day long.

The 1984 Carrera that followed is a further refinement of the 911SC. The 3.2-liter motor produced about 27 more horsepower, and the 1987 and later models feature the beefier G50 transmission. These are great cars, and nearly identical to earlier 911SCs. Now that both series are fairly old, the best car to buy is going to depend on an individual car's overall condition.

THE PEOPLE'S SPORTS CAR
CABRIOLET

From 1948 through 1965, Porsche cars were associated with open-air motoring. The very first Porsche was an open-top car. The first Porsche race car, the 550, was designed to be an open-aired, aluminum-shelled racer. The Porsche 356 lineup included a cabriolet until it was discontinued in 1965. So, when the 911 was introduced in 1965, nearly everyone anticipated that the Porsche 911 cabriolet would follow.

Alas, the 911 cabriolet did not materialize. Coincidentally, 1965 was the year Ralph Nader's ground-breaking book, *Unsafe at Any Speed*, was published. The book criticized auto manufacturers for their lack of attention to safety features, and targeted the design of convertibles. Convertibles had previously accounted for 6.3 percent of total US car sales, but that number dropped to about 1 percent in 1970. Elevated public awareness of automotive safety issues, and discussions of roof crush standards, decimated sales of open-topped cars. Porsche chose the safe path and further developed the Targa, which was a compromise between open-top motoring and safe structural protection.

Porsche in 1983 finally released a true convertible, the 911SC cabriolet, and sold more than 4,000 open-top units out of about 12,500 SCs produced. The car featured a manually folding top, meaning the driver had to park the car, unlatch the top, and then manually push it down over the rear seats. In 1986, Porsche offered a fully-powered, push-button top.

Modern-era Porsche 911 convertibles are designed with advanced features like solid-glass rear windows, hydraulic mechanisms, and anti-rollover bars that pop up like airbags in the event of a rollover. While many car manufacturers have designed folding convertible roofs that are solid, Porsche has continued to use an all-cloth convertible top. The cloth top folds easily and takes up significantly less space, which is important, considering the engine is located in the rear of the car.

THE PEOPLE'S SPORTS CAR
964/993 (1989–1998)

As the 1980s drew to a close, Porsche released its latest update to the 911—the Carrera C4, also known as the 964. The car was a true evolutionary step from the earlier 1974–1989 G-Series 911, as Porsche added modern styling and built the car featuring 85 percent completely new parts and designs. The result was the best 911 yet, with new features like optional four-wheel-drive, an optional four-speed automatic transmission, anti-lock brakes, coil springs, and a vastly improved 3.6-liter variant of the traditional Mezger engine. The 964 also introduced the first active aerodynamics with a motorized spoiler that rose up at speeds above 50 miles per hour. Despite all of these improvements, Porsche sales lagged in the early 1990s, and for the six years that the 964 was in production, Porsche only sold about 10,000 per year, making them among the rarest 911 models.

In 1995, Porsche revised the 911 again and released the 993, a model that many revere as the pinnacle of air-cooled Porsche 911 production. Once again, the model was redesigned, with more than 80 percent of the parts being carried over from the 964. Porsche stylists improved the 964's design, creating a look that has aged very well. The 993 featured a new multi-link suspension, a newly designed light-alloy chassis, and a small boost in horsepower mostly provided by a new VarioRam induction system. For many enthusiasts, the Porsche 993 represents an era of Porsche's past excellence, a time when the doors closed with a distinctive, high-quality "thunk," and nearly every element of the car feels evolutionary from the early 911s. The 993 is the last of the air-cooled Porsches, and values of these cars have increased significantly in recent years as buyers have realized that the car represents the best and final model from the "good ol' days" of Porsche.

THE PEOPLE'S SPORTS CAR
PETER SCHUTZ

FUN FACT

Peter Schutz passed away in late 2017, but left the world with a great gift, his book *The Driving Force: Extraordinary Results with Ordinary People*. It is chock-full of stories about his time at Porsche, and he self-narrated the audiobook.

HISTORICAL TIDBIT

With no chance for an overall win, Schutz vetoed running Porsche 944s at Le Mans in 1981. Instead, the company entered the Porsche 936 with an older "Indy" engine that was literally pulled from the Porsche museum to run at Le Mans, where it won. This same engine model was later massaged into the powerplant of the 956 and 962.

KEY PERSON

Helmuth Bott was the head of Porsche's research and development department in the 1980s. He spearheaded the design and production of the world's first supercar, the legendary Porsche 959.

Building tractors for Caterpillar and diesel engines for Cummins doesn't seem like it would qualify someone to run one of the world's premier sports car companies. But that's exactly the story behind Porsche CEO Peter Schutz. In 1980, Ferry Porsche personally asked Schutz to apply for the company's top position. Approved by the board of directors, Schutz became the first American to run Porsche.

Porsche underwent trying times in 1980. Sales were stagnating, the company had lost money for the first time in its history, and the future was very uncertain. The board voted to end production of the marque's flagship car, the Porsche 911. Three weeks into the job in early 1981, Schutz reversed that decision and refocused the company on the racing sports cars that had previously made it successful (see next section, "The Near Death of the 911").

Merely reviving the 911 would not be enough to turn around the company. However, instead of focusing on textbook, cookie-cutter lessons taught in business school, Schutz's first steps were to rebuild the company culture where all employees were considered part of a family striving for shared success. One pillar of this strategy was to expand and develop the racing program to win major races once again.

Schutz's strategy proved immensely successful. From his start at Porsche in 1981 through his retirement in 1987, a Porsche-powered race car won every 24 Hours of Le Mans, every 12 Hours of Sebring, and every 24 Hours of Daytona. It's hard to beat that record.

"Race on Sunday, sell on Monday" is an old auto industry phrase that encapsulates the theory that racing success corresponds to increased sales of production cars. Schutz understood this relationship, and during his tenure at Porsche, sales nearly doubled from 28,000 cars in 1980 to about 56,000 at the peak in 1986.

THE NEAR DEATH OF THE 911

In the trying year of 1980, then-Porsche-CEO Ernst Fuhrmann began to wind down 911 development and racing activity in favor of the 928. Ferry Porsche, chairman of the supervisory board, ousted Fuhrmann and took a chance on American Peter Schutz, who, as CEO, quickly moved to save the Porsche 911.

The story, as told by Schutz, is now a Porsche legend. On the job as CEO for merely three weeks, Schutz learned that the Porsche board had decided to kill off the Porsche 911 and replace it with the Porsche 944 and 928 lineup. Feedback from dealers had indicated that the 911 was expensive, lacked in quality, and handled poorly. Schutz instantly recognized that company morale was suffering because the 911 had come to symbolize the company's soul and essence, something the board failed to realize.

Schutz headed to the office of Porsche's chief engineer, Helmuth Bott, to discuss plans for upcoming models. While there, Schutz noticed a timeline chart on the wall that showed the 944 and 928 extending into the future, while 911 production ended abruptly in 1981. In a flash of American bravado, Schutz picked up a marker from Bott's desk and drew a line for the 911 off the chart, onto the wall, and out the door. When he re-entered the office, Bott was smiling. Schutz asked, "Do we understand each other?" and with a nod from Bott, the 911's future was restored.

A few years later, design work on the 959 was initiated. What started as a design project with almost no budget constraints soon defined an entirely new category; the 959 became known as the world's first supercar. This would not have happened if the 911 had been killed off in 1981.

956/962: Porsche's legendary race car from the 1980s, considered the most successful race car in history. Its results include seven victories at the 24 Hours of Le Mans, four victories at the 12 Hours of Sebring, and five victories at the 24 Hours of Daytona.

Airbox pop-off valve: An aftermarket valve installed in the 911SC airbox to prevent engine damage in the case of an accidental backfire.

Boxster: Porsche's two-seater roadster that was unveiled at the 1993 North American International Auto Show in Detroit. The Porsche 996 and Boxster share the same chassis/platform.

Cabriolet: A body style where the roof is retractable (also called a convertible).

Camshaft: Used to control the opening and closing of valves in the cylinder head.

Chain tensioners: Components in the engine that keep the timing chains tight when the engine is operating.

CIS: Continuous Injection System, a type of fuel injection developed by BOSCH and used on 911s from mid-1973 through 1983.

Cylinder: The engine piston rides up and down inside a cylinder.

Fuchs: Otto Fuchs Metall GmbH was the manufacturer of the distinctive "windmill" style aluminum alloy wheel commonly installed on the 911. This wheel was revolutionary due to its lightweight design and strong forged construction.

G50 Transmission: Porsche's replacement for the 915 transmission, installed in the Carrera beginning in 1987.

Galvanizing: A process of coating steel as a rust-prevention measure.

Generator: An early automotive component driven by the engine that generates electricity. Later cars have a similar component called an alternator.

Internal combustion engine: An engine that uses fuel combustion to create power.

Le Mans: A French city on the Sarthe River, and home to the 24 Hours of Le Mans race held each June.

Long hood: Early 911 models from 1964–1973 are typically referred to as "long hoods," largely because the hood design was shortened in 1974 due to new bumper regulations.

Mezger engine: The original Porsche 911 engine designed by Hans Mezger; variations of this design were used until 2012. It features a boxer piston layout and a single overhead camshaft on each side of the engine.

Mid-engine: A chassis design where the engine is located between the front and rear axles.

Push rods: Rods used to push open the valves in a cylinder head.

Targa: A body style invented by Porsche that resembles a convertible, but includes a large anti-rollover roll bar.

Tatra: A Czechoslovakian automobile and truck manufacturer.

Uni-body: A design for a car chassis where the body panels support the suspension points (as opposed to frame design, where the body shell sits on the frame).

VarioRam: A Porsche induction system that varies the length of the inlet ducting depending upon engine load and speed.

Volkswagen: A large German auto manufacturer founded in 1937.

Zündapp: A major German motorcycle manufacturer founded in 1917.

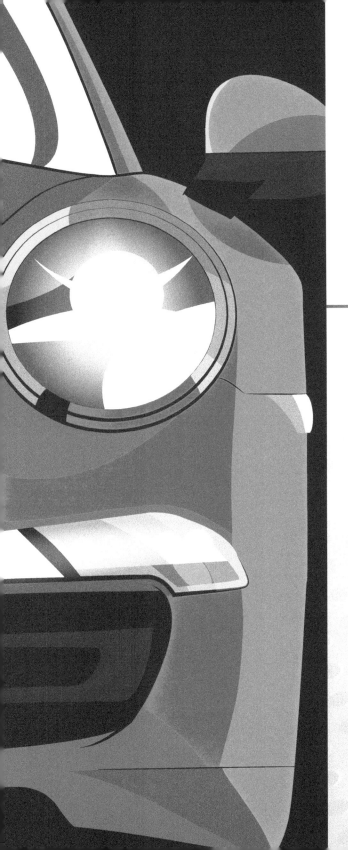

WASSERBOXER
A NEW KIND OF 911

WASSERBOXER
EMISSIONS REGULATIONS

In 1998, Porsche aficionados lamented the end of an era as the last of the air-cooled Porsches rolled off the line. With the end of 993 production and the introduction of the 996, the air-cooled era was over. While the Porsche 996 and its descendants are excellent sports cars, some in the Porsche world feel that a true 911 should always be air-cooled.

Air-cooled engines, as the name implies, use air to cool the cylinders and oil to cool the rest of the engine. Liquid-cooled engines use a water-based coolant to transport heat away from the entire engine, and the coolant then flows through an air-cooled radiator, typically located at the front of the car. While air-cooled systems are simpler and generally more reliable, water-cooled systems provide quicker warm-up times and more precise control of the operating temperatures within the engine.

The shift from air-cooled engines was driven by three factors: stricter emissions regulations, restrictions on engine noise, and the desire to squeeze more power out of the engine. For various technical reasons, air-cooled engines on average produce higher levels of pollutants than similar water-cooled engines. In addition, by the late 1990s, air-cooled cylinder heads had been massaged to be as large as possible. Due to design limitations of an air-cooled cylinder, the engine can typically only have two valves per cylinder. Liquid-cooled engines can generally have four valves per cylinder, which accommodates faster volumetric fuel flow and better performance and fuel economy.

Using liquid cooling, Porsche could squeeze more life out of the venerable Mezger engine design, a platform that had been in use since 1963. For Porsche's super-high-performance cars such as the 911 Turbo and the 911 GT3, the engineers adapted the time-tested Mezger-designed air-cooled motor to utilize water-cooled cylinders. The last year of the Mezger engine was 2012, with Porsche stretching the displacement to 4.0 liters and output to about 500 horsepower.

FUN FACT
The first water-cooled Porsche 911 was the Porsche 959 supercar. In a prophetic glance into the future, the 959 engine was designed with four-valve, water-cooled heads and air-cooled cylinders. This higher-horsepower-output engine required the four-valve spark plug in the center of the cylinder head design.

HISTORICAL TIDBIT
The liquid-cooled engine that replaced its air-cooled predecessor was called the M96. Although it certainly wasn't Porsche's first liquid-cooled motor, the M96 suffered from a host of design problems that included premature bearing wear, cracked cylinder heads, and leaky head gaskets. Later versions of the M96 were more durable, but they lacked the reliability of the 993 engine.

KEY PERSON
None. The M96 liquid-cooled engine had so many issues that no one wants to be associated with it.

WASSERBOXER
HARM LAGAAY

FUN FACT

The 1993 Porsche Boxster concept was a fabulously attractive car, but several features made it impossible to drive. The production version differed significantly from the show model, with the body modified to accommodate the wheels, suspension, and engine.

HISTORICAL TIDBIT

Harm Lagaay helped design the early Porsche 911 in the 1970s and designed the Porsche 924, a project originally intended to be developed for Audi. In the 1980s, he designed the BMW Z1 with its interesting "disappearing doors" and color-changeable body panels.

KEY PERSON

Grant Larson was the American designer of the original Boxster concept, and the winner of an internal race to design the production model. Larson's design evoked the past with lines drawn from the first Porsche racer, the 550 Spyder. Lagaay made the decision to use Larson's design in production.

The job of designing a new 911 is not one most people would enjoy. Not only does the design have to hew to the traditional silhouette of the 1963 original, it must incorporate innovative, forward-thinking features and elements.

This pressure is what confronted head designer Harm Lagaay when he took control of the Porsche Styling Department in 1989. The world was enduring an economic recession, Porsche sales were slowing, and the company's long-term survival was at stake. The 1989 introduction of the lightweight, nimble, and just-plain-fun Mazda Miata had re-ignited the small, two-seat roadster market, and Porsche took note. While the Miata had the edge in pricing and production volume, Porsche set out to develop the Boxster, a mid-engine sports car that could not only be the "best in breed" among two-seat sports cars, but would also provide the company with a versatile platform on which to base future models (such as the up-coming 911).

In a nutshell, Lagaay succeeded on all accounts. The stunning Boxster prototype was revealed in January 1993 at the Detroit Auto Show, earning high praise from the motoring press. The Boxster was rightly billed as revolutionary, and Porsche announced a month later that the concept would go into production.

When the "new" 911 was introduced in 1998, its front end was nearly identical to that of the Porsche Boxster. Porsche purists were not pleased, as it was very difficult to distinguish a Boxster from a 911 by viewing the front of the car. The purists rebelled, calling the 911 headlamps "fried eggs" and lamenting the way Porsche had simply cloned the front of the Boxster for the 911. Shortly before he retired, Lagaay corrected this issue with a facelift on the 2005 Porsche 997, returning the 911 to its circular headlamp roots, a design feature that remains in use today.

WASSERBOXER
THE BOXSTER SAVES THE 911 AND PORSCHE

FUN FACT

There was a running joke that one couldn't tell a Boxster from a 996-based 911 by only looking at the front end. (Porsche wisely changed this on later models.)

HISTORICAL TIDBIT

The Boxster's success could be traced to the Mazda Miata, which reenergized the modern two-seater roadster category when it was unveiled in 1989.

KEY PERSON

Wendeling Wiedeking was the CEO in charge of Porsche at the time, and made the difficult decision to bet the farm on the Boxster/996 platform. Without the Boxster, Porsche would likely have been an acquisition target, and the fate of the 911 and the entire marque—well who knows what would have happened . . .

The Boxster is arguably the most important car in Porsche's history. The Boxster?! Yes, the development of the Boxster turned Porsche around and allowed it to survive when the company's future didn't look too promising.

To fully understand its impact, one has to recall the early 1990s. Porsche struggled to sell cars through a recession that began in 1990. Porsche's lineup was stale, the 928 and 944 lines were faltering, and the future looked bleak. To put things in perspective, Porsche in 1993 sold only 3,700 cars in the United States. With about 300 US dealers at the time, that meant each dealer only sold about one car per month! Terrible indeed.

In a incredibly insightful move, Harm Lagaay urged Porsche management to green-light the Boxster concept into a production model. It became the high-performance car of choice in the modern roadster category, ultimately competing against rivals such as the BMW Z3, Audi TT, and Mercedes SLK.

So what does this have to do with the 911? Well, not only did Porsche design a new category killer in the Boxster, the company also adopted many design and manufacturing techniques learned from Japanese manufacturers—Toyota in particular. Utilizing common parts and similar designs let Porsche develop a platform on which multiple models could be built. Most people don't realize it, but the 911 released in 1999 is nearly identical to the Boxster in many ways. The models use shared designs for the electronics, braking system, engine, and transmission, and in the beginning, their front ends were virtually identical.

WASSERBOXER
WENDELIN WIEDEKING

If the Boxster deserves credit for saving the Porsche lineup, then Wendelin Wiedeking deserves credit for saving Porsche as a company. Wiedeking was the CEO of Porsche AG from 1993 through July 23, 2009, and he oversaw the company's regeneration and rebirth. Under his watch, Porsche went from selling about 3,700 cars in the United States to selling approximately 35,000 in 2007, right before the "great recession" hit.

Achieving this success was no small feat, and required significant forethought. Wiedeking wisely killed off the 928 and 968 lines, leaving Porsche with just one model, the 911. The company went all-in on the Boxster/996 platform, whose Japanese-influenced design and assembly techniques reflected Porsche's new focus on profitability and manufacturing efficiency.

Wiedeking was also instrumental in one of the most controversial decisions in company history: the development of the Porsche Cayenne Sport Utility Vehicle. Often panned by Porsche purists, the vehicle was undeniably a huge commercial success and the revenue it generated enabled Porsche to continue development of the Boxster and the 911. Built on the Volkswagen Touareg platform, the Cayenne risked being viewed as a rebadged Volkswagen—similar to what happened with the Porsche 914 and the Porsche 924 (which was originally designed to be an Audi). However, that did not happen, and the Cayenne proved capable of taking on the big boys like Range Rover and Jeep.

Wiedeking finished his career at Porsche assisting the Porsche family with an audacious gamble to acquire Volkswagen AG. This attempt ultimately failed and created the odd situation where acquirer became the acquired as Volkswagen ended up acquiring Porsche.

WASSERBOXER
996/997

FUN FACT

Due to the location of the center driveshaft on four-wheel-drive models, fuel levels on the 996 and 997 can only be measured from full to half-empty. Fuel levels from half-empty to empty are estimated by a computer analyzing fuel economy and driving style.

HISTORICAL TIDBIT

Of the seven generations of Porsche 911 created so far, the 997 has been the most popular, with 213,000 cars manufactured.

KEY PERSON

Grant Larson, designer of the 997, made the rear of the car wider, but the most significant change was the return of the oval headlamps traditionally used on earlier 911s.

The Porsche 996, revealed in 1999 as the Carrera 4, represented the biggest change in the Porsche 911 lineup since the model was introduced in 1963.

Porsche purists lament that the 996 represents the abandonment of air-cooled engines. To them, any definition of a 911 must include an air-cooled engine. In reality, the 996 was the best-performing 911 at the time, and represented a quantum leap in performance over the previous-generation 993.

But the 996 Achilles' heel was the new M96 engine. Designed for assembly efficiency, it seemed to be a misstep by Porsche. Although the engine maintained the "flat-six boxer" configuration synonymous with the Mezger engine, the new M96 engine was completely liquid-cooled, and did not have a separate dry sump system. This proved disastrous for racing, so these engines were never seriously used in competition except for light-duty club racing and track days. The design of the intermediate shaft bearing (known as the IMS bearing) proved to be another weak point that has caused many engine failures over the years. On most engines, the bearings can be retrofitted with more-durable aftermarket kits.

The 996 also suffered from an aesthetic issue: it too closely resembled the cheaper and slower Boxster. When the 996 made its debut in 1999, the 996 and Boxster front ends were virtually identical, and 911 buyers were upset they had paid twice the price for a car that looked just like a Boxster! Porsche "fixed" this issue in 2002 with redesigned headlamps on the 996, and again in 2005 when the 997 received "retro" circular headlamps that more closely resembled those of 911s produced prior to 1999.

WASSERBOXER
991

FUN FACT

The 991 features a small button mounted on the steering wheel called the "Porsche Sport Response Button," or PSRB. Push it and the car goes into attack mode for 20 seconds: the gearbox downshifts, the engine revs higher, and the turbo boost limiters are temporarily disabled, permitting overboost. It's the perfect feature for passing slow Priuses driving 54 miles per hour in the fast lane.

HISTORICAL TIDBIT

In 2014, Porsche released a striking redesign of the 911 Targa. The new Targa mixed the modern new design of the 991 with the old-school flavor of the classic Targa bar.

KEY PERSON

Michael Mauer was the chief designer when the Porsche 991 was developed. In 2012, the newly redesigned Porsche 911 won the "Red Dot" award for product design. Mauer is also responsible for the Porsche 918 Spyder design.

Released for the 2011 model year, the new Porsche 911—internally called the 991—was built on the success of the Porsche 997. The 991 was an entirely new, evolutionary platform designed to echo the spirit of all the previous models while delivering improved technology and higher performance.

The Porsche 991 integrated abundant new technology to the 911. In 2013, the GT3 was introduced with active rear steering for improved cornering. The PDK double-clutch transmission is an engineering marvel; essentially two transmissions in one that shift between each other effortlessly and with no perceptible loss of power. Electronically controlled differentials split the torque between wheels based upon vehicle speed and road conditions. Porsche Active Suspension Management (PASM) offers continuous adjustments to the damping force on each wheel based on driving style and current road conditions. Optional Porsche Carbon Ceramic Brakes (PCCB) help shorten braking distance and offer excellent fade resistance and incredibly light rotating mass. The all-new 9A1 engine was completely redesigned to eliminate issues that dogged the previous-generation M96 engine.

The second version of the 991 was revised for the 2017 model year. Following a trend among high-performance car manufacturers, Porsche downsized the engine and introduced twin-turbochargers as the standard engine configuration. Although critics noted this configuration was not traditional in the "Porsche 911" sense, the engine packages delivered improved performance.

Automotive critics and reviewers saw the 991 for what it truly was—a technological evolution of the Porsche 911 brand. The car is significantly larger than the 1964 original, as evidenced by Porsche's use of 19-inch and 20-inch wheels as standard equipment. It is indeed truly remarkable that the car, originally designed more than fifty years earlier, continues to evoke similar passion and exude proven performance.

CARRERA

Say the word "Carrera" around the globe and nearly everyone knows you're talking about a high-performance Porsche. Named after Porsche's early successes in the Carrera Panamericana race, the label was initially used by Porsche to denote its highest-performing models. The Porsche 356 Carrera was built with the four-cam, high-performance racing engine deployed in the 550 Spyder. The 904 Carrera GTS (Gran Turismo Sport) lived up to the name, and was positioned as a "race car for the street." The 906—or Carrera 6, as it was known in sales brochures—continued the tradition in 1966.

The name was shelved from 1966 until about 1973, when Porsche released the 911 Carrera RS, a model many regard as the pinnacle of development and the ultimate 911 from both historical and performance perspectives. In 1974, Porsche began to overuse the Carrera name. The 1974 Carrera models in the US only had minor improvements over standard 911s. This dilution continued in 1976 with the introduction of the Carrera 3.0 and Turbo Carrera, two standard-configuration models available in Europe. The Carrera name had come to mean, essentially, 911.

The trend continued with the introduction of the highly success-ful 911SC, with the SC standing for "Super Carrera." In 1984, confusion reigned again when the "Super Carrera" was replaced with the next-generation "Carrera" that was available from 1984 through 1989.

With the 1989 introduction of the 964, the Carrera name was often unofficially abbreviated in the form of C2 (two-wheel drive) and the C4 (four-wheel drive). With the introductions of the 996 and 997 in the 2000s, the confusion continued, and "Carrera" to this day is largely syn-onymous with nearly any Porsche 911.

991: The 991 is the seventh generation Porsche 911, released as a 2011 model.

996/997: The 996 and 997 are the fifth and sixth generations of Porsche 911. The 996 was introduced as a 1999 model, and the 997 was introduced as a 2005 model.

9A1 engine: Porsche's all-new engine introduced with the 991. This engine features numerous reliability improvements and direct fuel injection (DFI).

Air-cooled: An engine is air-cooled if it uses a fan to blow cool air across its cylinders. Air-cooled engines commonly use oil for cooling as well.

Cayenne: Porsche's highly successful Sport Utility Vehicle (SUV), unveiled in 2002.

Driveshaft: On four-wheel-drive Porsches, the driveshaft runs down the center of the chassis and delivers power to the front wheels.

Dry sump: Dry sump engines have a separate oil tank that holds a large quantity of oil in reserve. Dry sump engines are especially important in racing, where extreme operating conditions may cause oil starvation problems on traditional "wet sump" engines.

Emissions: The exhaust gases emitted from the engine as a byproduct of the combustion process.

Intermediate shaft bearing (IMS): The IMS bearing supports a shaft in the M96 engine that links the camshafts. This bearing has been a common source of failure in this engine.

M96 engine: The entirely new engine developed for the Porsche 996 platform. Although a high–performance powerplant, the M96 engine suffered from a variety of design/reliability issues.

Miata: Madza's ultra-successful roadster, also known as the MX-5. The Miata was launched in 1998 and reignited the two-seater roadster category.

Overboost: A condition that occurs when the turbos are allowed to boost air pressure beyond what would normally be acceptable pressures.

Porsche Active Suspension Management (PASM): Porsche's proprietary suspension control system that varies shock absorber dampening based upon driving style and road conditions.

Porsche Carbon Ceramic Brakes (PCCB): Porsche's proprietary ceramic brake system, featuring ultra-light brake discs that help minimize brake fading.

Porsche Sport Response Button (PSRB): This button places the car into an aggressive acceleration mode to maximize performance when overtaking slower traffic.

Roadster: A two-seater convertible.

Touareg: Volkswagen's SUV platform, upon which the Porsche Cayenne SUV is based.

Valves: The valves in a cylinder head open and close to allow fuel in and exhaust gases out.

Water-cooled: An engine is water-cooled if it uses a water-based coolant to remove heat from the cylinders, cylinder heads, and engine block.

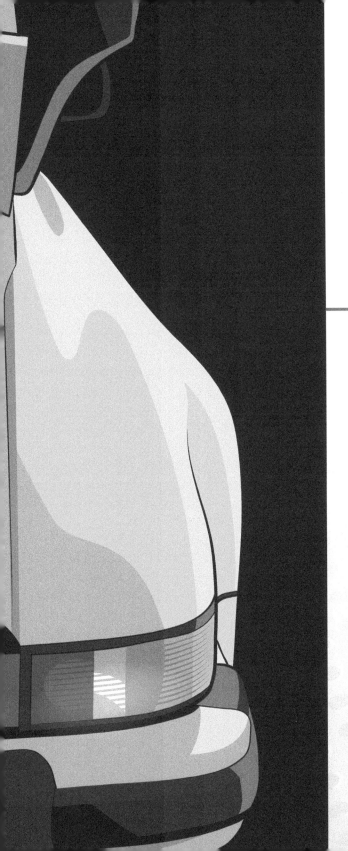

PUTTING THE CART BEFORE THE HORSE

ANATOMY OF A 911

PUTTING THE CART BEFORE THE HORSE
GERMAN ENGINEERING

FUN FACT

Bertha Benz, business partner and wife of automaker Karl Benz, took the first motorcar on the world's first road trip with her two teenage sons in August 1888. She reportedly did so without informing her husband. Serving as both driver and mechanic, she drove 121 miles and made many impromptu repairs along the way.

HISTORICAL TIDBIT

The roots of German engineering can be traced back to the time of the Vikings. Very successful "German-engineered" products from the time were Frankish swords traded throughout Europe under the brand name Ulfberht. Their metallurgy was considered superior and they are jokingly referred to as the world's first branded luxury products.

KEY PERSON

Karl Friedrich Benz was a German engineer widely credited with the design and production of the world's first automobile, the 1885 Benz Patent-Motorwagen.

In modern times, the phrase, "German engineered" has become synonymous with the concepts of unique and forward-thinking design, creative use, and premium product quality. In Germany, the prevailing thought is that if you purchase an item, you want to own it for a long time, and you want to appreciate its design, feel, and construction while you use it. Whether it's a writing pen or bullet train, the design philosophy is the same, and it's ingrained in the German culture. Germans are taught from an early age to design for excellence, not to cut corners, and to seek perfection.

The German educational system is also very focused on vocational training. Secondary school students are assessed and tested to determine which students should attend college. Those not considered to be ideal college candidates are offered an educational track through company apprenticeships combined with vocational training in a related field. This dual-track method helps ensure universities have a ready supply of well-focused students, while companies hiring apprentices acquire workers familiar with the company's practices, insights, and processes—inside and out. This dual-path approach to education is considered a key pillar in Germany's rise as a modern industrial power.

German engineering is about designing for purpose, quality, and longevity. The fruits of this mindset can be seen in Porsche's racing successes. Time and time again, Porsche has won races not necessarily by being the fastest, but by being very fast *and* reliable. Countless times in racing history, competitors have designed cars to beat Porsche, and they do succeed—for a while. However, grueling races like the 24 Hours of Le Mans or the Paris-Dakar rally require reliability as much as speed, and Porsche has earned its reputation for engineering durability with many come-from-behind victories in races where the front-runners just couldn't last.

PUTTING THE CART BEFORE THE HORSE
REAR ENGINE

The 911 has its engine located in the rear—meaning that it's located behind the rear axles. There are a few advantages to a rear-engine design.

First, the design places the majority of the engine's weight under the rear wheels. Since most rear-engine cars also use rear-wheel drive, this design increases the downward force on the rear wheels, resulting in greater traction on slippery surfaces. Secondly, rear-engine design affords car designers ample room up front for passengers and luggage. Since the transmission is typically a combination transaxle (a transmission and rear differential combined), the resulting powerplant is generally very compact.

In addition, the engine and transmission can be easily packaged together and installed in the car in a relatively simple manner. This aids in assembly and maintenance. Braking performance is also improved. Under heavy braking, the front end of a car typically "dives" down due to weight transfer and the increased load on the front suspension. Rear-engine cars are more balanced under braking because there is less weight in the front. Also, under acceleration, more weight is transferred to the rear of the car, which is desirable for increased traction.

What are the downsides to a rear-engine design? With all of that weight in the rear, the 911 has a reputation for oversteering. This is the tendency of a car to pull sideways through corners, with the back end tending to swing out wide, possibly causing the car to spin. This is most commonly experienced on cars with worn or cold tires, or on slippery surfaces such as ice and snow.

PUTTING THE CART BEFORE THE HORSE
AIR COOLING

FUN FACT

The fan belt is one of the most critical components in an air-cooled engine. Porsche 959 and 962 race car engines have a pressure sensor in the fan's airflow, and if the fan belt breaks, a warning lamp lights up on the dash.

HISTORICAL TIDBIT

In 1975, Porsche had a problem meeting stricter US emissions standards with its air-cooled engines. Porsche switched from an eleven-blade fan to a five-blade fan from 1975–77, which resulted in higher engine temperatures and reliability issues, but let the company achieve legal emissions levels.

KEY PERSON

Hans Mezger was the designer of the original air-cooled 911 engine and the twelve-cylinder air-cooled engine that powered the Le Mans—winning Porsche 917.

The concept of air-cooled engines has been a hallmark of the Porsche 911 from day one, and its roots stretch back to the Porsche 356 and Volkswagen Beetle, both of which were air-cooled.

The term "air-cooled" is a bit of a misnomer. The 911 engine is indeed air-cooled, but the cooling airflow directed to the engine is used to cool only the cylinders and cylinder heads. The rest of the engine is cooled by oil. The engine incorporates a dry-sump system in which a large external oil tank supplies oil to the engine. A larger oil supply removes heat from the engine's long block, then flows through one or more oil coolers.

One signature component of the 911 is the large, multi-blade, horizontally mounted fan that faces the rear of the car. The fan is as well-known a symbol as the 911's silhouette, and is even featured on numerous T-shirts. The fan directs air down across the heads and cylinders and out the bottom of the engine compartment. Nearly all air-cooled 911 owners understand the importance of the fan's rubber belt; if the belt snaps or slips off, cylinder temperatures will spike and the resulting heat will fry the engine.

Air-cooled engines have heat exchangers that heat air that is then cycled into the passenger cabin. A trademark smell of the Porsche 911 is that "air-cooled smell," which is really just motor oil that has seeped onto the heat exchangers. The fumes from this burned oil mix with fresh air and flow into the passenger compartment. Even fixing every oil leak leaves the faint oil smell in the system, and most 911 owners can identify the scent in a fraction of a second!

PUTTING THE CART BEFORE THE HORSE
HANS MEZGER

All-star engineers don't have their pictures on bubble gum cards, but if they did, Hans Mezger would be the engineering equivalent of Babe Ruth. Mezger joined Porsche after completing college in 1956 and went straight to work in Porsche's Works 1 development department. Over the next thirty-five years, he was the driving force behind Porsche's legendary engines, including the original flat-six 911 engine, the twelve-cylinder 917 engine, and the legendary TAG turbo engine that powered McLaren to multiple championships in the 1980s.

The four-cylinder 356 engine was directly inspired by the Volkswagen air-cooled engine, and was Porsche's go-to production engine for decades. Porsche wanted a new, larger, higher-performance engine for the 911, one that improved on the reliability of the older push-rod design. Using lessons learned from developing Porsche's 753 and 771 race engines, Mezger designed a new air-cooled engine that incorporated a dry-sump system for a more reliable oil supply, as well as a chain-driven single overhead cam. These improvements were considered necessary for racing and performance driving, and the design was so successful, it remained largely unchanged until it was retired in 2013.

The 911 engine repeatedly proved itself to be reliable and successful on the track, so Porsche decided to build higher-horsepower variants. For the 908 engine, Mezger extended the engine case, crank, and cams to incorporate an additional cylinder, and he added two gear-driven camshafts to each bank of cylinders. The results were expectedly successful, with the 908 winning a total of five times at the Nürburgring Nordschleife.

By 1970, Porsche needed an even larger engine to compete against the twelve-cylinder Ferraris. Mezger applied the same trick again, but this time essentially combing two motors together to produce the twelve-cylinder 917 engine, which powered winners at Le Mans in 1970 and 1971.

ADDITIONAL READING: *Porsche and Me*, by Hans Mezger with Peter Morgan

PUTTING THE CART BEFORE THE HORSE
SUSPENSION

The uniqueness of the 911 extends to its suspension design. With the engine weight in the rear of the car, Porsche engineers had to design a suspension system that would counter the inherent oversteer and the car's tendency to spin out in turns.

The front suspension has traditional A-arms coupled to torsion bars that act as springs. The rear wheels are hung on semi-trailing arms that are sprung using horizontally mounted torsion bars. The rear torsion bars eliminate the need for springs to be wrapped around the shocks.

When the 911 wheelbase was increased in 1969, the rear trailing arms were lengthened by 61mm. This resulted in a near-identical increase in the car's wheelbase (which helped reduce the rearward weight bias). This was the most significant 911 suspension change until the 964 was introduced in 1989.

The 964 four-wheel-drive system didn't have room for the torsion bars, so Porsche moved to a coil-over front suspension design. The rear suspension utilized coil-over springs as well, along with a pivoting link in place of the inner arm, which helped reduce the rear-engine oversteer effect.

In recent years, Porsche engineers developed Porsche Active Suspension Management (PASM), which offers continuous adjustments to the damping force on each wheel based on driving style and current road conditions. In addition, Porsche Dynamic Chassis Control (PDCC) was developed as an electronically controlled anti-roll bar system that works with PASM to limit side-to-side chassis roll in response to road conditions and driving style. With the release of the 918 Spyder, Porsche developed rear-wheel steering, which turns the rear wheels in the opposite direction of the front wheels at low speeds. This increases maneuverability and assists in reducing the oversteer that is inherent in the rear-engine design.

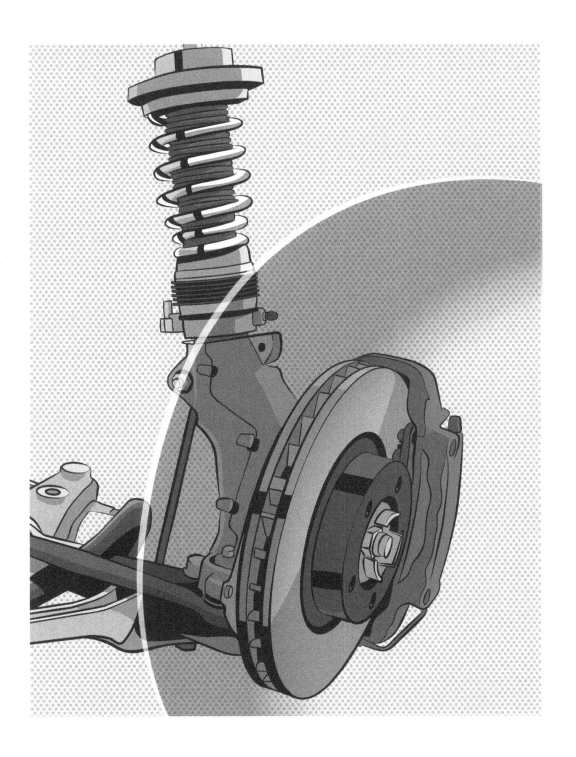

PUTTING THE CART BEFORE THE HORSE
AERODYNAMICS AND THE SILHOUETTE OF THE 911

It's rare that the US Trademark and Patent Office will allow the specific trademark for the shape of a car, but that occurred with the Porsche 911. The 911 silhouette is so unique and instantly recognizable worldwide that the shape has become a brand of its own. It's also highly unique in the automotive industry in that it has maintained its distinctive silhouette largely unchanged for many decades.

The initial 911 design process was filled with strife and conflict. Ferry Porsche, the head of the company at the time, enlisted his son Ferdinand Alexander Porsche (also known as "F. A." or "Butzi") to help sketch the new design. Butzi's designs were well received by everyone except Erwin Komenda, head of the design department and designer of the Porsche 356. After some internal conflict, Butzi's designs were taken across the street to the coachbuilder Reutter. The shape developed in the Reutter shop became the basis for the prototype revealed at the 1963 Frankfurt Motor Show.

The new design paid tribute to the 356 that had launched the company, yet it also incorporated several improvements over the 356 that Ferry Porsche had desired. Ferry wanted the new car to provide more interior space and comfort than the 356, and Butzi's design did just that. It incorporated traditional round headlamps into the two freestanding front wings, and the trunk had a more angular design than the rounded 356. The fastback design was extended and smoothed out when compared to the 356, and the side windows were stretched to fit the silhouette's free-flow lines. When compared side-by-side, the family resemblance of the 911 and 356 is obvious, but the refinements are truly evolutionary and they live on in modern 911s today.

PUTTING THE CART BEFORE THE HORSE
TURBO

Turbocharging has been synonymous with Porsche performance for several decades. When Hans Mezger designed and developed the Porsche 917 in the late 1960s, there were limited restrictions and regulations on engine size. Porsche's domination with the 917 in the early 1970s led to rule changes and reclassifications that limited engine displacement. Porsche's response was to turbocharge smaller-displacement engines.

How exactly do turbos make more power? A turbocharged engine has some "assistance" in filling the combustion chamber with the air/fuel mixture. On a normally aspirated engine, the maximum manifold pressure is atmospheric pressure (14.7 psi). On a forced-induction engine, manifold pressure is increased by the turbocharger. An increased mass of air and fuel is injected into the combustion chamber, resulting in more power. The turbocharger compressor is driven by excess engine exhaust, and although the backpressure on the exhaust may rob a small amount of power from the engine, the turbo boost is generally thought of as free boost. Typical peak revolutions of turbos can range from 75,000 all the way up to 150,000.

Porsche's first major turbocharging effort came with the repurposing of the 917 for the Can-Am series. Adding a pair of massive turbochargers to the 5.4-liter engine produced a reliable 1,100 horsepower, with up to 1,500 horsepower available in qualifying rounds. Porsche also turbocharged the 936, an open-cockpit prototype racer that laid the foundation for the 956/962.

The results of Porsche's efforts speak for themselves. Of eighteen overall wins at Le Mans, sixteen of those were turbocharged Porsches. Similar results were achieved at Daytona, Sebring, and numerous other circuits. The turbocharged 956 and 962 became the most successful race car platform ever, with more than 140 wins over more than a decade of racing.

PUTTING THE CART BEFORE THE HORSE
ALL-WHEEL DRIVE

When Porsche developed the 959 in the mid 1980s, it challenged its engineers to create the best sports car they could—with almost no budget limitations. They were challenged to develop systems that didn't exist and create technologies that hadn't been invented yet. One of the advances they developed on the 959 was a form of advanced computer-controlled four-wheel-drive.

The 959 four-wheel-drive system was complicated and unique in design. The car's power is distributed between the front and rear wheels using a computer-controlled multi-plate hydraulic clutch located in the center of the car. Normally set to a 40 percent front, 60 percent rear power split, the computer adjusts these levels based upon inputs from the car, as well as a driver-controlled setting (for normal, rain, snow, or off-road conditions). The 959 drive system was expensive and complicated, and although it performed well, it was viewed as overkill for a standard production car.

The 964 Carrera 4, released in 1989, used a much simpler four-wheel-drive system. Three differentials use computer-controlled hydraulic clutches to maintain the full-time all-wheel-drive system's power distribution at about 31 percent front and 69 percent rear. Called the Porsche Dynamic All-Wheel Drive System (PDAS), it was very capable yet relatively basic by today's standards. The 993 drive system was simplified even further, implementing a limited-slip differential in the rear combined with a viscous coupling located in the center of the car.

What does all this mean for performance? While most people want all-wheel drive for driving in snow, Porsche designed its AWD systems for higher performance on the racetrack. When a traditional rear-wheel-drive car corners, the weight transfer tends to lift the inside wheel, resulting in reduced power to the outside wheel, particularly if the car doesn't have a limited-slip differential. On a four-wheel-drive car, power is routed to the front inside wheel in cornering, allowing maximum power grip to the road.

PUTTING THE CART BEFORE THE HORSE
WATER COOLING

For decades, the Porsche 911 was known as an air-cooled car. That was part of its legacy, its heritage, its culture. To many, a 911 that was not air-cooled was blasphemy. Alas, late-1990s regulations that called for reduced emissions dictated that air-cooled cars were soon to become a thing of the past.

Air-cooled engines typically run a slightly richer air/fuel mixture than water-cooled engines, which decreases fuel efficiency and increases exhaust emissions. In addition, air-cooled engines usually have only two valves per head, largely due to the fact that the head needs to include cooling fins. Water-cooled engines can incorporate a center-plug, four-valve design that is generally more efficient at fuel flow and ignition.

A big step in the transition to water cooling came with the 1978 935/78, which featured a new head design bolted onto the standard 911 long block (the "Mezger Engine"). With four valves per water-cooled head, this engine was used in the ultra-successful Porsche 956/962, the Porsche Supercar 959, and the 911 GT1, along with the modern-day GT2, GT3, and 996/997 Turbo.

Unfortunately, when Porsche designed the Boxster and 996 in the late 1990s, it used a completely different engine design, the M96 motor. While a fine engine, this powerplant was not reliable enough for serious racing. Designed for manufacturing efficiency instead of racing durability, the M96 engine slightly tarnished the 911's reputation for "track-ready" performance on the street. Instead of the M96, most modern 911 race cars use a variant of the original "Mezger engine" used in the 956/962.

ANACHRONISM OR TECHNOLOGICAL MARVEL?

The Porsche 911 has always been known as a quirky car. Things are just different on the 911—not necessarily better or worse, but different. The ignition key is on the left side of the steering wheel. The engine is in the back and is air-cooled. The suspension uses torsion bars and spring arms. The door locks require you to turn a knob to open them, and the windshield wipers are on the left side of the car. The engine has a separate oil tank, and checking the oil level requires a degree in engineering. The heating system is almost always terrible, as oil leaks from the engine onto the heat exchangers and creates that distinctive "air-cooled smell" in the passenger cabin that some 911 aficionados simply adore. The sound of the air-cooled Porsche 911 engine is distinctive and can be instantly recognized by nearly any early 911 owner.

What about the newer cars? The shift began in 1999 with the retirement of the air-cooled engine and the introduction of the new, modern-era 911s built using lean manufacturing techniques and with a water-cooled engine installed. The purists will argue that the last "true" 911 was the 1998 air-cooled 911; however the past twenty years have proven that the new models are just as capable of maintaining the "911 legend" as the older ones.

Indeed, the future lies with the new owners. As the 996 and 997 cars become less expensive, a breed of new, younger owners are buying the cars, repairing them, and modifying them in their garages. So far, the 997 has been the most popular, with 213,000 units manufactured. Porsche has made more than a million 911s and more than 700,000 of them are still on the road. Porsche serves not only as the manufacturer but also the steward of the Porsche 911 brand, and takes great pains to protect the image, feel, smell, and spirit of the 911 in each generation it has built.

GLOSSARY: PUTTING THE CART BEFORE THE HORSE

908/3: Porsche's eight-cylinder racing car, successfully campaigned at the Targa Florio race.

917: The Porsche 917 is a twelve-cylinder race car that won Le Mans overall in 1970 and 1971, Porsche's first overall Le Mans victories.

918 Spyder: Porsche's hybrid gasoline-electric supercar from 2013.

959: Porsche's supercar project from the 1980s. The 959 featured the best technology at the time and essentially created the supercar category.

962: The most successful race car of all time, Porsche's dominating ground-effects Le Mans racer from the 1980s. The 962 was a slightly more developed version of its predecessor, the 956.

A-arms: Suspension components that support the front spindles and travel up and down as the car rides over bumps in the road.

Boost: The amount of increased intake-air compression gained from a turbocharger.

Heat exchangers: Used to transfer heat from the exhaust to the air that will be circulated to the passenger compartment for heating purposes.

Normally aspirated: An engine with an air intake that is not turbocharged and not supercharged.

Oversteer: When going into a corner at speed, a car with oversteer characteristics tends to want to spin out (common with rear-engine cars).

Porsche Dynamic All-Wheel Drive System (PDAS): A full-time all-wheel-drive system used on the 964 that delivers power at ratios of about 31 percent to the front wheels and 69 percent to the rear wheels.

Porsche Dynamic Chassis Control (PDCC): PDCC was developed as an electronically controlled anti-roll bar system that works with PASM to limit side-to-side chassis roll in response to road conditions and driving style.

Rear-engine: When a car's engine is located behind the rear axles.

Torsion bars: A type of torsional spring used to spring-load the suspension of the 911 (1964–89).

Trailing arm: A suspension design used on the independent rear suspension of the 911 (1964–98).

Turbocharging: Turbocharged cars use exhaust gases to compress the air/fuel mixture and increase the total amount of air and fuel injected into the engine to increase power output.

Understeer: When going into a corner at speed, the weight from a front-engine layout will tend to make a car with understeer characteristics push into the corner and resist turning.

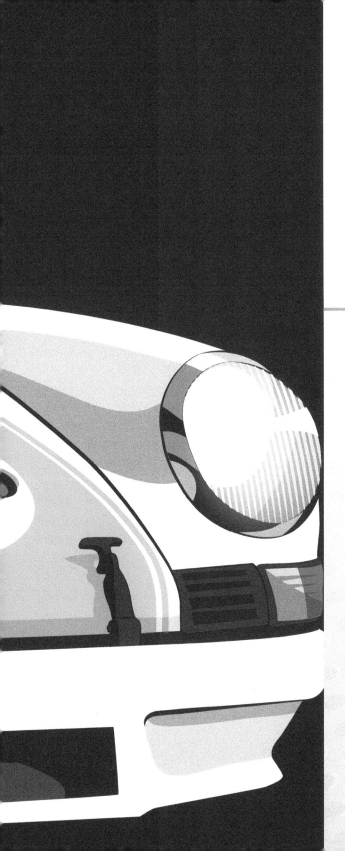

DOING IT
IN THE ROAD
ROAD RACING THE 911

DOING IT IN THE ROAD
ROAD RACING

The Porsche 911 is unique to the world of motorsport, not so much for its number of victories, but more for how its capabilities inspired Porsche to develop some of the world's greatest race cars. The 911's racing history really took off in 1967 with the lightweight 911 R—one of Porsche's first successful attempts at racing its "newly designed" car. In the mid-to-late 1970s, it was the Porsche 935, a 911-based, 800-horsepower beast that dominated the competition. Finally, Porsche's 962 (1982–1994), the most successful race car of all time, is largely based on technology first deployed in the Porsche 911. For example, the 962 engine uses the same engine-case casting that was employed in the 911SC.

The 911 has long held a well-earned reputation as a street-legal car that can be driven to the track, raced, and then driven home. In fact, that is exactly what the factory did in 1967 when it needed to send a replacement car from Germany to the racetrack at Monza for an endurance racing time trial. Even today it's not uncommon for racers to drive their 911s to local club events—sometimes with track tires mounted on the roof!

Modern-day 911s have continuously proved their racing chops as well, as shown by the success of the Porsche Cup and GT3 race cars at both European and American circuits. The popularity of 911 GT3 racing in Grand Am and the American Le Mans series reinforces Porsche's reputation for designing versatile, customer-friendly race cars. In addition, Porsche's on-track support ensures that its customer race teams are well equipped and prepared at every event.

DOING IT IN THE ROAD
911 R

"Lightweight, nimble, and quick" describes most early Porsche sports cars. From the fiberglass 904 Carrera GTS to the lightweight Speedster Carrera and even the aluminum 550 Spyder, by 1967 Porsche had developed a playbook of race car designs that proved successful both on the track and in the dealerships. The 911 R was born out of that experience.

For the development of the 911 R, Porsche basically took a base 911 and lightened it in the extreme—even replacing the metal door handles with ones crafted from plastic. The results yielded a car that weighed a mere 800 kilograms (1,763 pounds). The engine was a racing version of the standard 911 flat-six, similar to the one installed in the Porsche 906 race car. With approximately 210 horsepower, the 911 R's power-to-weight ratio in kilograms was an impressive one-to-four.

Due to its limited production run, the 911 R was raced by the factory in the 2-liter GT prototype class (GTP) where it finished third behind two Porsche 910s at the Circuito del Mugello. The 911 R also performed well in endurance racing, winning the 84-hour Marathon de la Route on the Nürburgring with Vic Elford and Jochen Neerpasch behind the wheel. The 911 R was also an effective rally car, winning such races as the Rallye Coupe des Alpes in 1967—again with Vic Elford driving.

A total of nineteen 911 R cars were built, fifteen of which were sold to the public, with the remaining four retained as factory race cars. The 911 R's rarity combined with its distinction as the "first 911 race car" makes it perhaps the most valuable and sought-after 911 ever made.

ADDITIONAL READING: *Porsche 911 R-RS-RSR*, by John Starkey

DOING IT IN THE ROAD
911 RSR/934

The 1973 911 Carrera RS was produced by Porsche in order to homologate the model for racing in the FIA Group 4 category. The Carrera RS was designed and built for the street, and the RSR was the full-blown racing version.

Porsche manufactured about sixty RSR cars, internally known as the M491 option. Compared to the RS, these cars featured wider fender flares; brakes developed for the 917; and a slightly larger 2.8-liter, 300-plus horsepower race engine running a 10.3:1 compression ratio, with twin-plug ignition, revving to 8,000 rpm. The car featured lightweight doors and decklids to help drop its weight to 1,958 pounds. RSRs featured 9-inch-wide wheels at the front and 11-inch wheels in the rear.

The RSR made its mark quickly, winning its debut race—the 1973 24 Hours of Daytona—with Peter Gregg and Hurley Haywood driving. Gregg and Haywood followed Daytona with a victory at the 12 Hours of Sebring. A 911 RSR also won the 1973 Targa Florio, and the combined victories earned Porsche the World Manufacturers and IMSA Drivers championships that year.

The 24 Hours of Le Mans race proved another matter. The 911 RSR was outclassed by lighter and more powerful prototype cars like the Ferrari 312P and the Matra MS670, with the Martini RSR managing fourth at the end of the race. While the RSR dominated the GT class in 1973, the mismatch at Le Mans set the stage for Porsche's development of the Carrera RSR Turbo (1974) and the turbocharged 934 (1976).

Based on the production 1976 Porsche 911 Turbo, the 934 was armed with a 480-horsepower engine and a top speed of 190 miles per hour. It earned Porsche several victories in the 1976–77 seasons, including the European GT Championship and the Trans-Am championship. More importantly, the 934 was the steppingstone and development platform for the 935, which would dominate racing in the following years.

ADDITIONAL READING: *Carrera RS*, by Thomas Gruber and Georg Konradsheim

DOING IT IN THE ROAD
935

FUN FACT
The original Transformers cartoon featured a Porsche 935 as the character Jazz, the right-hand robot to chief robot Optimus Prime.

HISTORICAL TIDBIT
Computer giant Apple has only sponsored one race car in the history of the company— a Porsche 935. This car won both Daytona and Sebring and recently sold for almost $5 million.

KEY PERSON
Norbert Singer, considered the father of the 956 and 962, honed his skills with the development of the mighty 935.

Porsche is renowned for carefully studying the racing rules and maximizing any advantage it can find through loopholes and exceptions. The 935 is historical proof of that approach.

The Fédération Internationale de l'Automobile (FIA) rules for Group 5 specified that the hood, roof, doors, and elements of the roof line were to be left unmodified—everything else was up for grabs. What Porsche designed looked like a cartoon caricature of a 911. The rules didn't mention a required location for the headlamps, so Porsche flattened the nose of the 911 and placed the headlamps down low, just above road level. The wheel arches were enlarged, exploiting a loophole created by a requirement that said only that the doors had to be in the same location as on a production car. Further modifications to the floors allowed Porsche to replace much of the chassis with tubular frame—bolting in the roof, window pillars, and a production floor to follow regulations.

Kremer Racing, following Porsche's lead, came up with its own unique modification. The rules mandated that the rear window and roof must remain in the same place as on the stock production cars. Kremer cunningly designed a second complete roof and window that fit over the stock one, resulting in a much smoother and more aerodynamic profile. Deployed on their 1979 Le Mans entry, this trick helped Kremer win the overall victory.

The super-reliable Porsche 935 dominated racing worldwide from 1976 through the mid-1980s, earning overall victory at Le Mans in 1979, six wins at the 24 Hours of Daytona (1978, 1979, 1980, 1981, 1982, 1983), and six wins at the 12 Hours of Sebring (1978, 1979, 1980, 1981, 1982, 1984). The only force able to stop the 935's racing dominance was the alteration of the racing categories by the FIA. Porsche responded with the 956 and 962, which continued dominating racing for another decade.

ADDITIONAL READING: *Porsche 930 to 935: The Turbo Porsches,* by John Starkey

DOING IT IN THE ROAD
PETER GREGG, IMSA CHAMPION

Peter Gregg was a true Porsche enthusiast, racing a 904 and 906, and in 1965 purchasing a Porsche dealership in Jacksonville, Florida, named Brumos Porsche. In the 1970s, the Brumos livery of red and blue stripes on white race cars became synonymous with Porsche racing victories in America.

Gregg claimed several victories in the Trans-Am Series and started off the 1973 season teamed up with co-driver Hurley Haywood. The two co-drove a 1973 911 Carrera RSR for the overall victory at the 24 Hours of Daytona, then went on to win seven more races that year in the RSR— forever mating the Brumos name and livery with the ultra-successful Porsche 911 RSR. So unstoppable were Gregg and Haywood that the duo were soon nicknamed "Batman and Robin" by the motorsport press.

Gregg posted a total of sixty-two victories over his career, along with 116 podium finishes. His six IMSA overall championships in the 1970s, combined with an unparalleled attention to detail, earned Gregg the nickname "Peter Perfect," a reference to the cartoon character of the same name from the late 1960s television show *Wacky Races*.

In 1980, Brumos built a new Porsche 935 with improved aerodynamics that was intended to replace Gregg's championship 1979 935. Unfortunately, Gregg suffered a brain injury in an off-track car accident and could not manage to qualify on the new car's debut at the Daytona Finale. Suffering double vision and barred from racing by IMSA, Gregg tragically took his own life soon thereafter.

ADDITIONAL READING: *Hurley from the Beginning*, by Hurley Haywood and Sean Cridland

DOING IT IN THE ROAD
911 GT1

The mid-1990s saw a revival of sports car racing with the FIA GT Championship. Porsche's previous efforts in sports car racing had produced the 962, which to this day remains the most successful race car of all time. For the GT Championship, Porsche brought to bear its rule-bending skills and designed the 911 GT1, effectively a prototype sports car racer. They also designed a street-legal version to comply with homologation requirements: the 911 GT1 Straßenversion, a stunning street-legal Le Mans prototype racing car.

The first generation 911 GT1 basically melded a 993-generation front chassis with a rear engine and suspension derived from the racing 962. Porsche won the GT1 class at Le Mans in its debut race, but lost the overall victory to Joest Racing's Porsche-powered WSC-95 prototype.

In 1998, Porsche completely redesigned the GT1. With an all-new carbon fiber chassis, and a newly designed sequential gearbox, the car featured entirely new bodywork that looked less like a street car and more like a sports-prototype race car of the era. This new 911 GT1-98 struggled throughout most of the 1998 FIA International GT season, with the notable exception of that year's Le Mans race. The GT1-98 leveraged Porsche's proven reliability and notched first and second place finishes at Le Mans after BMW, Mercedes, and Toyota suffered mechanical problems. The Porsche factory poster celebrating the win proclaimed, "Surviving the 24 Hours of Le Mans takes a very good car. Winning, on the other hand, requires a Porsche."

Later that year, in one of the most spectacular crashes in racing history, a factory Porsche GT1 racing at Road Atlanta Petit Le Mans was caught in the draft of the car in front of it. Air trapped beneath the GT1 caused it to perform an almost perfect summersault flip, lifting the nose and flipping 360 degrees through the air only to land on the rear two wheels. Miraculously, four-time Le Mans winner Yannick Dalmas survived the real-life rollercoaster ride with only a few minor bumps.

ADDITIONAL READING: *Der Erfolgstyp Porsche 911 GT1*, three-book series by Ulrich Upietz

DOING IT IN THE ROAD
911 GT3 RS/SUPERCUP

With the development of the 911 GT3 and its associated racing series, Porsche completely redefined motorsports racing over the past twenty-five years. In 1993, Porsche developed the Porsche SuperCup—a Porsche 911–only racing series in which all the cars were similarly equipped, driven by privateer teams, and incredibly competitive. The SuperCup series and GT3 Cup Challenge series were a throwback to true "stock car" racing, when the cars were based on road-going vehicles and were nearly identical in performance.

What makes the series so popular with enthusiasts? Perhaps it's the thrill of racing when all drivers are on equal footing—eager newbie drivers can compete alongside veteran professionals. Because the cars are all so similar, minor tweaks and adjustments in the pits can mean the difference between winning and last place. Porsche Motorsports has done a marvelous job promoting the series and supporting the cars at all the events, making participation relatively easy and relatively inexpensive (especially when compared to F1).

The 911 GT3 is the true star of the series, and Porsche offers turn-key cars for sale from the factory, complete with spares packages. Privateers can purchase either a street-legal version of the GT3 or acquire one of the numerous "slightly used" GT3 race cars that are retired each year. The cars are so plentiful that often older ones can be purchased for much less than the price of a new street-legal 911.

All of the 911 GT3 cars from 1999 through 2013 utilized a variant of the Mezger engine as their primary powerplant—a much-refined version of the same engine that basically started it all in 1963. The last version of the Mezger engine in 2012 was a 4.0-liter monster that offered 493 horsepower—an impressive 123 horsepower per liter for a normally aspirated engine. In 2013, Porsche introduced an entirely new, 475 horsepower 3.8-liter direct fuel injection (DFI) engine for the GT3.

HURLEY HAYWOOD, DAYTONA CHAMPION

Porsche has had plenty of successful race drivers, but one would be hard-pressed to name one more victorious than Hurley Haywood. He won Le Mans three times, all in a Porsche. He won the 24 Hours of Daytona five times, all in a Porsche. He won the 12 Hours of Sebring twice—both times in a Porsche. To say that Haywood and Porsche racing are near synonymous would be an understatement.

Haywood got a major break at age twenty when he beat Porsche dealer and race car driver Peter Gregg in a local autocross. Hurley and Gregg became friends and driving partners thereafter, famously winning the 1973 24 Hours of Daytona in a Carrera RSR 2.8. Haywood and Gregg would continue their friendship—sometimes smooth, sometimes rocky—until Gregg's unfortunate suicide in 1980. Haywood eventually became a Brumos vice president and would help lead the dealership and racing program in Gregg's absence.

Haywood has credited a lot of his success to having really great cars. Loyal to Porsche, Haywood only strayed to the Jaguar team when he injured his leg and couldn't summon the stamina to repeatedly engage his Porsche's clutch—something that wasn't an issue in the Group 44 Jaguar XJR-5.

Though Haywood achieved tremendous success Porsche's 935, he saw the future in the 956/962. In perhaps the most memorable Le Mans finish in history, he helped drive the Rothmans 956-003 to an amazing finish in 1983. The car overheated and broke down on the final lap of the race, but co-driver Al Holbert was able to restart the engine and limp the smoking 956 across the finish line a mere 17 seconds ahead of Derek Bell closing fast in yet another Rothmans 956.

In 2009, at the age of sixty, Haywood returned to the podium at Daytona for the fifth time, placing third with a Brumos-built prototype racer powered by a 4.0-liter Porsche engine (Brumos teammates won the race in a similar car). Haywood officially retired from racing in 2012, and he has just completed work on a new book and movie highlighting his career with Porsche and Brumos Racing.

GLOSSARY: DOING IT IN THE ROAD

Brumos: A well-known Porsche dealership located in Jacksonville, Florida, at one time owned by the late Porsche race driver Peter Gregg.

Compression ratio: The ratio of the maximum to minimum volume in the cylinder of an internal combustion engine. A high compression ratio typically translates into higher performance.

FIA: Fédération Internationale de l'Automobile—the governing body of motor sport that sets rules to promote safety and classifications for various auto racing events around the world.

Grand-Am and American Le Mans: Two independent sports car racing series that were merged into one in 2014.

Group 4/Group 5: Two classifications for motorsport racing, regulated by FIA. Group 4 race cars had more similarities to actual production cars than Group 5 race cars.

GTP: Gran Turismo Prototype. A classification of racing in the IMSA GT series.

GTS: Gran Turismo Sport, the name originally given to the Porsche 904 GTS. Originally the GTS classification meant additional comfort in an aggressive sports car.

Homologation: Most commonly used to refer to the number of production vehicles that must be built and sold to the public in order to qualify a racing version of the car for competition.

IMSA: The International Motor Sports Association, a North American racing sanctioning organization founded by John Bishop in 1969 and focused on road racing.

Nürburgring: A motorsports complex located in Western Germany that is home to a 12.9-mile-long loop track. It is often used by Porsche as a proving ground.

Podium finish: When a driver finishes a race in first, second, or third place.

Power-to-weight ratio: The ratio of horsepower to the weight of the car. A higher ratio means that the car will be faster on the track.

RS and RSR: Abbreviation for Rennsport, which means "racing sport" in German. RSR means "racing sport race car" in German (rennwagen).

Sequential transmission: A type of transmission that shifts gears one at a time in order, up or down, from a gear-shift lever. Allows for quicker and more accurate shifts.

Street-legal: Describes a car fitted with all the state- and federally mandated equipment for normal road use. Dedicated race cars often have much of this equipment removed for weight savings.

Twin-plug ignition: The use of two spark plugs per cylinder on an internal combustion engine. The use of twin-plug ignition allows engines to be operate at higher than standard compression ratios.

Weissach: A town in Germany, home to Porsche's research center and test track.

DOING IT
IN THE DIRT
RALLY RACING THE 911

DOING IT IN THE DIRT
RALLY RACING

For those who think regular racing on pavement is boring, someone crazy and thrill-seeking decided to invent rally racing. In a nutshell, a rally is a race contested off-road and on public roads with highly modified produc-tion cars designed for speed and off-road capabilities. Early rallies were based upon time trials, stages raced by a driver and a navigator with points earned for timekeeping and navigational accuracy with recorded times be-tween specified points.

In the 1950s, Swedish and Finnish competitors took the sport to a new level with the addition of special stages that were scored primarily on the speed in which they were completed. This led to the modern era's elabo-rate race setups on public and private roads with large corporate sponsors and races spanning several continents. In addition, races were often run in harsh conditions with events spanning deserts, snow-covered roads, and dirt/gravel roads.

In the 1980s, Audi developed the Quattro, a revolutionary all-wheel drive car that significantly raised the bar for off-road performance. Few other manufacturers had four-wheel-drive production cars to compete with the Quattro, so FISA (Fédération Internationale du Sport Automobile, the sanctioning body) opened up the competitions to Group B race cars. Thus began the legendary development of all-wheel-drive cars like the Renault 5 Turbo, Ford RS200, Peugeot 205 T16, Lancia Delta S4, and of course the legendary Porsche 959.

Porsche has achieved great success in rallying, but its results in road racing at Daytona, Sebring, and Le Mans tend to overshadow the rally races it ran and the records it set. Porsche's rally cars have maintained the marque's reputation for durable cars that can excel in off-road racing. From the 911's first major race at Monte Carlo to the 959's victory across the Saharan desert, the 911 has continually defined itself as a hearty and victorious rally competitor.

DOING IT IN THE DIRT
MONTE CARLO/VIC ELFORD

What do you get when multiple rally teams race from spots across Europe to Monte Carlo? The Rallye Monte Carlo. The historic rally begins at cities across the continent and runs on challenging roads to the finish at Monte Carlo. The most exciting section is the timed run from La Bollène-Vésubie to Sospel over the Col de Turini pass road, which includes hairpin turns, narrow cliff roads with drop-offs, and often challenging weather conditions.

The Porsche 911—born to be raced—was entered, shortly after its introduction, in the 1965 Monte Carlo Rally, where it finished fifth. This early success from what was basically a production car (albeit piloted by Herbert Linge and Porsche development engineer Peter Falk) proved the Porsche 911 had a promising future in rally racing.

The rear-engine design proved beneficial as the car's weight was significantly biased toward the rear. More weight on the rear wheels results in better traction and reduces the need for four-wheel drive. Interestingly, the 1965 car had a rear-mounted shelf with leather hand straps at the rear of the car, above the engine grill, where the navigator could stand to add more rear weight and increase rear-wheel traction.

Porsche truly established itself at Monte Carlo in 1968 with four powerful weapons: legendary drivers Vic Elford and Pauli Toivonen and two rally-prepped 911 T cars. Elford piloted his lightweight, performance-modified, ultra-reliable 911 T to the overall victory, with Toivonen only about a minute behind in second. Porsche matched this 1-2 finish in both 1969 and 1970, cementing the 911's reputation as a worthy rally competitor.

DOING IT IN THE DIRT
911 SAFARI RALLY

FUN FACT

The African Safari Classic Rally is an open road rally, meaning the racing sections are not closed off to the public, so drivers must contend with goats, wandering villagers, and assorted wildlife as they power through the jungle.

HISTORICAL TIDBIT

In 1972, Porsche lent Sobieslaw Zasada a 911 S and a support mechanic. Assisted by a volunteer crew, he finished second overall. The car was not highly modified and was a true testament to the durability of the Porsche 911.

KEY PERSON

Jürgen Barth was in charge of Porsche's racing efforts in the early 1970s. An engineer, driver, and mechanic, Barth earned a reputation as one of the most versatile figures in German racing history.

A Porsche has competed in the East African Safari Classic Rally, often called the world's toughest rally, nearly every year since the 1970s. The Safari Rally is infamous for its arduous conditions, rapidly changing weather patterns, and insanely rough roads that make maintenance of the cars very difficult. In addition, intense heat and humidity wreaked havoc on cars, drivers, and mechanics.

In 1971, Porsche had a strong chance to win the International Rally Championship of Manufacturers over mighty Ford. Porsche entered three cars in the Safari Rally with Björn Waldegård (winner of the Monte Carlo Rally in 1969 and 1970), Åke Anderson, and Polish rally driver Sobieslaw Zasada at the wheels. Accompanied on the rally by Porsche technical engineer Jürgen Barth, each car was modified with a 2.3-liter flat-six engine with twin-plug ignition and a heavy-duty clutch.

The cars were strengthened with double-thick floorpans and extra sheet metal in various areas. Steel plates covered the oil tank and rear spring plates, and the cars were fitted with well-used, settled-in torsion bars since new ones tended to sag too much with all of the abuse from racing. Spare oil lines were plumbed for redundancy, and a completely separate fuel system with its own 20-liter tank was fitted under the front hood. One of the cars' sponsors was Sears Roebuck & Co., and the cars ran Sears tires (manufactured by Michelin) with distinctive whitewall stripes.

Unfortunately, all this preparation did not pay off. Waldegård made a tactical error trying to overtake Zasada in the dust and ran his car off the road. Zasada finished fifth with lessons learned for the next year (he took second place in 1972).

DOING IT IN THE DIRT
PARIS-DAKAR RALLY

FUN FACT

The Paris-Dakar rally has a racing class for support vehicles. All the teams have supply trucks that chase the race cars across the same terrain, although at a much slower pace, and compete in their own truck class.

HISTORICAL TIDBIT

Porsche built only twenty examples of the lightweight, durable 911 SC/RS, which proved very successful in rally racing. Designed and built primarily for Group B competition, they were used as a stopgap measure until the 959 was completed.

KEY PERSON

French motorcycle racer Thierry Sabine got lost while competing in the 1977 Abidjan-Nice Rally, and after realizing the challenge of racing across desert sand dunes, he organized the inaugural rally from Paris to the Senegal capital of Dakar.

The Paris-Dakar Rally is an annual rally that was originally run from Paris, France, to Dakar, Senegal. This off-road endurance race featured terrain that was much tougher than usual, with many sections of the race featuring sand dunes, rocks, mud, water, and other terrain challenges. Race stages of about 500 miles a day added up to the full 6,200-mile racecourse.

Porsche's first entry in the Paris-Dakar Rally resulted in victory in 1984, when René Metge and Dominique Lemoyne won the car class with a Rothmans-sponsored race version of the Porsche 911 called the 911 SC/RS or 953. The car was a highly modified four-wheel-drive version of the 911 Turbo chassis with a 3.0-liter engine running mechanical fuel injection.

In 1985, Porsche decided to prep and run the 959 in the Paris-Dakar Rally. The new 962-based Porsche 959 engine was not yet ready for production, so Porsche installed a semi-standard, 3.2-liter, normally aspirated engine in the 959s instead. The relatively untested cars performed poorly, with all three failing to finish.

In 1986, Porsche ran three fully race-equipped Porsche 959 cars in the rally. Not wanting to take chances, Porsche teamed two-time Dakar winner René Metge with Jacky Ickx for the race. Impressively, Porsche's two Rothmans-sponsored cars finished first and second. An interesting footnote is the third 959 Porsche entered. It was run as a support vehicle carrying tools and equipment for the other two entries, and it was just as reliable, quick, and competitive as the other two, finishing sixth overall. It likely would have finished third if it hadn't been tasked with hauling spare tires and equipment!

DOING IT IN THE DIRT
PIKES PEAK AND JEFF ZWART

FUN FACT

In 2007, Zwart teamed up with Porsche Cars North America to race a team of three Cayennes in the TransSyberia Rally from Moscow to Ulaanbaatar, Mongolia.

HISTORICAL TIDBIT

In addition to the Pikes Peak International Hill Climb, Zwart also ran the US Pro Rally Championship in an ANDIAL-built Carrera 4, scoring six victories in the 1993 and 1994 seasons. Zwart raced in the 1997 Panama-Alaska Rally in his Porsche 914-6, and won his class in the 2004 Baja 1000.

KEY PERSONS

In 2017, Zwart and teammate Cameron Healy finished third in the NORRA Mexican 1000 in a 911. It was the first time a production-based Porsche had finished a 1,000-mile Baja race in its fifty-year history.

Pikes Peak, a 14,115-foot-tall mountain in the Colorado Rockies, is home to the second-oldest race in America and one of the world's most unique motorsport events. The annual Pikes Peak International Hill Climb consists of a 12.42-mile track that includes 156 turns while climbing more than 4,700 feet from start to finish. The grades average about 7 percent, and the thin air of the high-altitude track creates unique challenges for the cars and competitors alike.

Racer Jeff Zwart has made the Pikes Peak event his unofficial calling card for racing and rallying, and his first victory in a Porsche came in 1994 when he won the Open Class Championship. Over a period of sixteen years, he drove twelve different Porsches and won eight Pikes Peak class championships. In 2010, when the course was still a mix of dirt and pavement (it is now completely paved), Zwart won the Time Attack Class in a Porsche 911 GT3 Cup car.

His most noteworthy success came in 2011, when Zwart drove a street-legal Porsche 911 GT2 RS the 1,132 miles from Porsche Motorsport in Southern California to compete in the Pikes Peak race. Porsche kept all of the creature comforts like air conditioning and GPS in the car but also installed safety modifications such as a roll cage, fire suppression system, race seat, and safety harnesses. Sponsorship stickers were applied and Zwart set a new Production Car record, covering the course a full 24 seconds faster than his 2011 time in a GT3 Cup car.

The 1970s were a golden era for sports car racing, with the Porsche 917, Ferrari 512, and Porsche 935 showcased as pinnacles of performance. However, toward the end of the decade, car manufacturers felt oppressed by racing's rules and red tape, and expressed concern that Group 4 and Group 5 racing was becoming too expensive. As a result, FISA created three new classifications—A, B, and C—with Group C cars including full prototype racers like the venerable Porsche 962, and Group A, a production-car-based group, designed to ensure a significant number of privateer teams could compete.

Group B was host to supercar versions of production cars. The intent was to encourage more manufacturers to participate in road and rally racing by accelerating race car development periods, increasing event publicity, and reducing the homologation requirements. The new rules meant that only 200 production units would be required, which was half the number of the previous Group 4. This let manufacturers race their highest-performance cars with significantly lower production costs.

The Group B era featured some of the world's most impressive sports cars—cars that might not have been developed without the class's less-restrictive rules. Based loosely on production platforms, these cars launched the supercar concept for both road racing and rallying. Group B cars included the Porsche 959, Ferrari 288 GTO, Audi Sport Quattro, Ford RS 200, Lancia Delta S4, and about fifty other cars scheduled for production.

Unfortunately, after a series of fatal accidents and crashes in Group B racing, FIA banned Group B cars in 1987. Group B racing had also proved to be very expensive for private teams. A 961 (the racing version of the 959) cost $325,000 at the time, and for that price, a team could acquire a non-factory chassis 962 and achieve better performance.

GLOSSARY: DOING IT IN THE DIRT

911 SC/RS (953): Porsche's first Group B rally car, loosely based upon the production car at the time, the 911SC.

911 T: Porsche's basic, introductory model of the 911.

961: Porsche's racing version of the 959. Only one car was produced; it is currently housed in the Porsche museum in Stuttgart.

ANDIAL: An engine builder formerly located in Southern California, and former home of Porsche Motorsports North America.

Daytona: A racetrack (Daytona International Speedway) located in Florida, and home to the 24 Hours of Daytona. The track is characterized by steep, 31-degree banking.

Fire suppression system: An internal fire extinguishing system required on all race cars.

FISA: Fédération Internationale du Sport Automobile, the governing organization for motorsports.

Four-wheel drive: When all four wheels of a car have power applied to them.

Group B: A classification for rally cars with almost no restrictions. Group B was banned after a series of accidents, some of which included fatalities.

Mechanical fuel injection: A type of fuel injection used on the production 911 in the early 1970s, and also extensively used in Porsche racing cars.

Monte Carlo: The largest city in the Principality of Monaco, which is located on the Mediterranean Sea between France and Italy. It is the end point of the Monte Carlo Rally.

Navigator: The companion/passenger in a rally car who is responsible for calling out to the driver all of the race directions, turns, and obstacles.

Off-road: Racing on unpaved dirt, snow, or desert surfaces.

Paris-Dakar: An intense rally originally run across the European and African continents that included significant driving across desert terrain.

Pikes Peak: A mountain in Colorado that is home to the annual Pikes Peak International Hill Climb.

Quattro: Audi's groundbreaking four-wheel-drive car from 1980.

Rally racing: A mixture of road racing and off-road racing, often held on public roads in multiple stages.

Roll cage: A set of interlocking metal support tubes that provide protection for drivers and passengers in case of a rollover.

Sebring: A racetrack (Sebring International Raceway) in Florida that hosts the 12 Hours of Sebring. The raceway shares some of its track with the local regional airport.

Spring plates: The rear trailing arms of the 911 suspension are sprung by spring plates attached to torsion bars.

Stage: Rally racing is divided into stages, each of which is timed.

WRC: World Rally Championship, a rally series organized by the FIA that culminates in a championship driver and manufacturer.

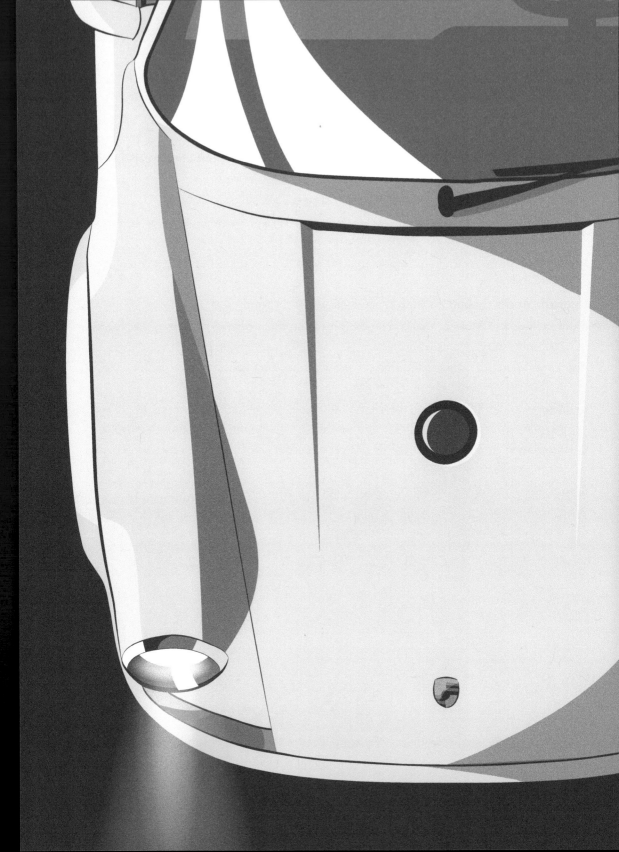

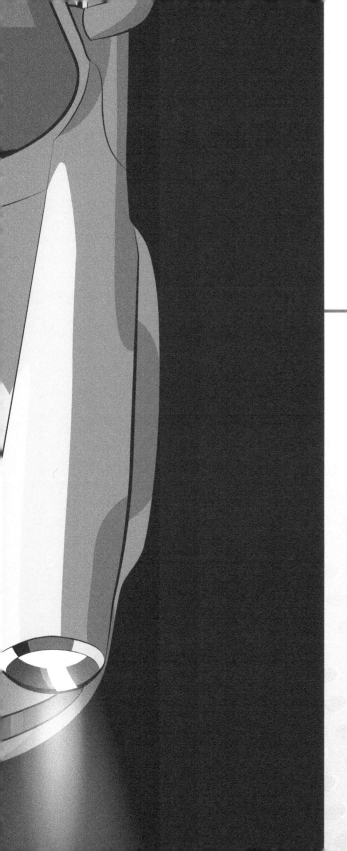

SPECIAL EDITIONS & TUNER CARS

THE ULTIMATE 911S

SPECIAL EDITIONS & TUNER CARS
1973 911 CARRERA RS

FUN FACT

Porsche wanted to make the Carrera RS as light as possible, so no rust-prevention coating was used on the chassis. As a result, one of the best ways to detect an "RS fake" is to check inside the engine compartment, where the sides of the sheet metal will be smooth and lacking any rustproofing materials.

HISTORICAL TIDBIT

The first 500 or so cars were manufactured with lightweight sheet metal. When Porsche extended the production run, standard-gauge sheet metal was used for the remainder of the production run.

KEY PERSON

Ernst Fuhrmann pushed for the concept of developing the Carrera RS instead of a 2.7-liter 911 S.

In 1972, Porsche had a problem. The world governing body for motor racing, the FIA, altered the rules to make the dominant Porsche 917 ineligible. The 911, with an engine displacement of less than 3.0 liters, became the focus of the Porsche racing department, and the legendary Porsche 911 Carrera RS was born (RS stands for *Rennsport*, or racin, in German).

To qualify to race in the Group 4 category (special GT), Porsche had to build 500 identical cars. The Carrera RS was made with lightweight sheet metal, a Spartan interior, and was armed with a 2.7-liter version of the 911 S engine. The car was a success on the racetrack and in the showroom, and it further cemented the 911 as an icon in modern car culture. The car featured the distinctive ducktail atop the engine lid, and it came with slightly wider rear wheels and fenders.

The Carrera RS was available in two distinct trims: lightweight and touring. The touring version closely resembled the trim of the 1973 911 S, with full "S" trim on the front and rear bumpers and the side rocker panels. The lightweight version featured fiberglass front and rear bumpers and no side trim pieces. Both versions sported the unique rear ducktail and the distinctive *Carrera* script name adorned both sides of the car. The engine was built with Nikasil-lined cylinders, was tuned to provide 210 horsepower, and utilized Bosch mechanical fuel injection (MFI).

The Porsche marketing department initially fretted about being able to sell all 500 cars, but the enthusiasm from the public was tremendous and Porsche actually produced a total of 1,580 units. In 1974, new laws mandated larger bumpers, and a less-powerful CIS fuel-injection system, making the 1973 Carrera RS the de facto pinnacle of early Porsche 911 development.

ADDITIONAL READING: *Carrera RS*, by Thomas Gruber and Georg Konradsheim

FLAT NOSE TURBO

FUN FACT

In the 1980s, several body shops and tuners converted regular 911s into Flat Nose versions. How to identify a fake? Nearly every factory version has a single headlamp door motor in the center of the trunk, with a set of rods that activate each headlamp door. Conversions typically have two motors, one for each headlamp.

HISTORICAL TIDBIT

In 1989, Porsche was pondering the end of the Turbo and decided to produce the 930 LE (limited edition) as a last hurrah. The 930 LE was basically a higher-performance, customized, non-slanted version of the Slant Nose Turbo. With only fifty produced, it is one of the rarest production Porsches.

KEY PERSON

Erwin Kremer, owner of the successful Le Mans–winning Kremer Racing team, helped popularize the conversion of street cars with Slant Nose styling that resembled the team's Porsche 935 race cars.

The 1980s produced some memorable trends, including big hair, Members Only jackets, Swatch watches—and the Porsche 911 Flat Nose Turbo. The origins of the Flat Nose (Flachbau) are rumored to trace back to 1980, when one of Porsche's race-team customers asked for a road car that resembled the Porsche 935 race car. Designed, engineered, and hand-crafted through Porsche's Special Wishes (Sonderwünsche) program, the Flat Nose—or slant nose—Turbo featured a prominently sloped front fender with flush-mounted retractable headlamps. Just ahead of the rear wheels were air vents, reminiscent of those of 1980s sports cars that provided additional cooling for the engine and rear brakes.

The Sonderwünsche program modified cars as a special customer-request option until about 1987, when it became an official option that dealers could order as option package M505 (USA) and M506 (the rest of the world). The price was $29,000 in 1980s dollars, which added about 50 percent to the price of a standard 911 Turbo! Approximately 900 cars were produced, making the Flachbau option fairly rare and highly desirable among collectors who appreciate the audacious styling.

While Porsche's Flat Nose option was the de facto standard body upgrade for the 911 Turbo, several other companies produced similar versions. Tuner company DP Motorsport designed and manufactured the Porsche 935 bodywork for Kremer Racing and built a handful of street cars based on its design.

An even rarer bird is the 1994 964 Turbo Flat Nose. Combining styling cues from the front of the Porsche 968 with rear vents inspired by the Porsche 959, the Flachbau X84 body package option for the 964 was offered near the end of the model's production run. Offered along with the "Turbo S" upgrade option (X88), only seventy-six examples were produced, making it one of the rarest production Porsches.

SPECIAL EDITIONS & TUNER CARS
THE SINGER

In 2009, rock musician and industrial designer Rob Dickinson came up with a crazy idea. It was inspired by the Southern California R Gruppe (Porsche 911 owners who like to tirelessly hot rod their roadgoing cars with sports/racing modifications) and by his experience modifying his 1969 911. Rob decided he would pursue the ultimate air-cooled evolution. Not unlike the factory-built Porsche 959 from twenty years earlier, there would be no compromises on performance, build quality, style, or budget. Every component was optimized—or, as Rob likes to say, "reimagined"—using the latest in modern technology. Retaining the essence and feel of the air-cooled 911s was essential.

The resulting car set a new standard that even Porsche finds difficult to match. Known as a Porsche 911 Reimagined by Singer Vehicle Design, each car is a bespoke client commission based on the 1990–1994 964. Each features carbon fiber bodywork, hand-woven quilted leather interior, bespoke seats, carbon fiber brakes, fully adjustable suspension, a completely new and modernized electrical system, and bespoke HID headlamps. Multiple engine options are available, including an Ed Pink–built 4.0-liter motor making 390 horsepower. The most extreme evolution is an all-new 4.0-liter, 500 horsepower, air-cooled flat-six designed by Williams in consultation with Hans Mezger, the designer of the original Porsche flat-six engine. This engine features a unique four-valve cylinder air-cooled head.

Details of what makes the Singer special are too numerous to list. What makes the cars unique and legendary is the fact that every element of the car is driven by Rob Dickinson's furious attention to detail, all the while keeping the passion, essence, and energy of the original air-cooled platform.

ADDITIONAL READING: *One More Than 10: Singer and the Porsche 911,* by Michael Harley and Rob Dickinson

SPECIAL EDITIONS & TUNER CARS
ENGINEERING MARVEL: PORSCHE 959

FUN FACT

The 959 has a very unique gauge as part of its five-gauge set. The four-wheel drive display shows the rear axle slip limiter on the left, and the percent of torque distributed to the front axle on the right. Four individual lamps in the center indicate which traction program is selected. In traditional 911 style, the 959 gauge has analog needles on a black background.

HISTORICAL TIDBIT

Porsche originally intended to produce and sell race versions of the 959 to their customers. Calling it the 961, Porsche produced one car, chassis 10016, that campaigned in a handful of races from 1986–87. The car met with lukewarm success, and when the Group B class was cancelled, Porsche had no championship series to run it in. Today, the lone car sits in the Porsche museum.

KEY PERSON

Professor Helmuth Bott was the chief engineer responsible for the development of the 959.

What do you get when you lock fifty or so German engineers in a room and ask them to develop "the ultimate sports car" with no budget limitations? That's exactly what Porsche did in 1984, and the result is widely considered the world's first supercar. The Porsche 959 redefined the sports car market at the very high end where there are no compromises on performance and no limitations on cost.

The Porsche 959 incorporates body panels manufactured out of light-weight aluminum and carbon-Kevlar. A total of thirteen computers control and coordinate every aspect of the car and are integrated through a central control unit. The 959 engine is a close derivative of the race engine used in the Porsche 962 and features the world's only sequential twin-turbo system, designed with both low-rpm and high-rpm turbos that work together to reduce turbo lag. The suspension system has dual shocks (one air and one hydraulic) on all four corners, which allows for adjustment between a sporty or comfort ride, as well as ride-height adjustment while driving. The viscous-clutch four-wheel-drive system utilizes multiple inputs to a computer that then distributes power to each wheel based upon acceleration and road conditions.

Porsche worked with Bosch to develop the world's first tire-pressure monitoring system, integrated with the design and fabrication of hollow-spoke wheels from Speedline. Porsche also worked with WABCO Westinghouse to develop the four-wheel anti-lock braking system that works in concert with the variable four-wheel drive.

Although Porsche only built a handful of 959s when compared to the regular production 911, the build-quality and design were performed as if the Porsche were going to produce hundreds of thousands of them. Porsche supposedly lost a fortune on the 300 or so 959s they built, but much of the technology developed for the 959 was also deployed in later versions of the 911—spring-boarding the 911 platform to a new level of performance.

ADDITIONAL READING: *Porsche 959*, by Jurgen Lewandowski

SPECIAL EDITIONS & TUNER CARS
A CLASSIC REINCARNATED: 911 SPEEDSTER

In 1954, American Porsche importer Max Hoffman dreamed up a budget-minded experiment that became the iconic Porsche 356 Speedster. Designed to be lightweight and cheap, the bargain basement Porsche Speedster ironically now sits at the top of the Porsche world in terms of value and desirability. All 356 Speedsters were convertibles, and the low, raked windscreen (removable for weekend racing), was a defining feature that helped contribute to the car's long-lasting popularity. It was dubbed the "bathtub Porsche" since it looked like an old-school bathtub turned upside down.

In 1989, with the Porsche 911 G-Series (1974–1989) nearing its end and sales of the 911 beginning to slow, Porsche decided to capitalize on the legend of the Porsche Speedster and build a 911 version (dubbed the M503 option). While the original 1950s 356 Speedster was born of a bare-bones approach, Porsche designed the 1989 Porsche 911 Speedster to be an expensive, limited-edition model.

The 911 Speedster convertible borrowed a few elements from its earlier brethren—most notably, the raked windscreen. Adorning the rear convertible storage area was a fiberglass cover dubbed the "Speedster humps." Nearly all of the 2,100 or so cars produced were manufactured with the "Turbo Look" package that consisted of extra-wide wheels and fenders borrowed from the 911 Turbo. Although the "Turbo Look" package portrays a unique and aggressive stance, the wider wheels and suspension from the higher-horsepower Turbo tend to make these cars slightly less nimble than the narrow-bodied ones.

Some consider the 1989 Porsche 911 as the last of the great 1980s Porsches, and the Porsche 911 Speedster signifies the pinnacle of that series. Porsche repeated the process to produce Speedsters in 1992, 2010, and 2019, but the 1989 version remains the most sought-after among collectible Porsches.

SPECIAL EDITIONS & TUNER CARS
THE WORLD'S FASTEST CAR: RUF & THE YELLOW BIRD

FUN FACT

Not all of the cars were yellow—at least one was painted mint green.

HISTORICAL TIDBIT

In 1984, the RUF BTR (Gruppe B Turbo RUF) broke the 300 km/h mark during testing for the *Road & Track* "World's Fastest Cars" issue. Lessons learned from the BTR, which was built on a classic 1978–1989 Porsche 911 platform, paved the way for the CTR.

KEY PERSON

Alois Ruf Jr. is the president and owner of RUF Automobile and lead engineer of the Yellow Bird.

Headquartered in Pfaffenhausen, Germany, RUF Automobile is known as one of the world's premier tuners of Porsche automobiles. RUF put its name on the world stage in 1987 when Alois Ruf produced a highly upgraded version of the 911 called the RUF CTR. Adorned in bright yellow paint, the RUF CTR was also known as the Yellow Bird and, in 1987, Phil Hill set the record for the world's fastest production car with a top speed of 211 miles per hour in a Yellow Bird. Starting with a standard 911 shell chassis and powered by a Porsche-based 3.37-liter turbo-charged flat-six engine, the Yellow Bird took the base Porsche 911 and just made everything better.

RUF paid attention to the small details like the removal of rain gutters and the welding up of seams, improved aerodynamics with the use of customized spoilers and air dams, and focused attention on removing excess weight. The resulting car was faster than the factory's best effort, the 959, posting a 0 to 60 (0 to 97 km/h) time of 3.65 seconds in testing by *Autocar* magazine. RUF designed distinctive 17-inch wheels (produced by Speedline) and mated them with Dunlop's Denloc run-flat tires (originally designed for the 959). The bored-out twin-turbo engine used an engine management system borrowed from the Porsche 962 and produced about 470 horsepower. Not happy with using the heavy-duty Porsche Turbo four-speed gearbox, RUF designed a new five-speed unit for the CTR and fine-tuned the gear ratios to achieve blistering top speeds.

It was officially named the CTR ("Group C Turbo RUF"), but the writers at *Road & Track* magazine nicknamed it Yellow Bird—and the name stuck. Its price was $223,000 USD, just $2,000 less than a base Porsche 959 at the time. (Both cars would sell for about $500,000 in inflation-adjusted 2018 dollars.) Officially, only twenty-eight of these monster cars were produced, but in later years RUF continued to convert customer cars to CTR-specifications.

SPECIAL EDITIONS & TUNER CARS
RAUH-WELT BEGRIFF

Porsche 911 owners typically fit into two distinct categories. One group feels every decision by the company's German engineers is sacred and unquestioned. The other group feels the Stuttgart engineers have designed a good platform suitable for modification. The early 2000s brought the movie franchise "The Fast and the Furious" to the silver screen, and with it an introduction to the Japanese culture of wild modifications to stock cars. The Japanese tuner leading the way with Porsche 911 modifications is RAUH-Welt Begriff (RWB).

RAUH-Welt translates to "Rough World," a name coined by the company's founder, tuner Akira Nakai, commonly known as Nakai-San. The cars are often modified from stock Porsche 993s, and are known as RWBs. Signature elements include absurdly large wheels with low-profile tires, exaggerated wheel arches riveted in place, and multi-level wings atop the engine lid.

RWB modifications and styling are polarizing, and have fans and detractors. But undisputed is RAUH-Welt's reputation for high-quality craftsmanship. Most of the changes are cosmetic, but many cars also get suspension upgrades. Much like Duesenbergs of yesteryear, each RWB is handcrafted and custom ordered, so each one is unique.

As one might imagine, Nakai's creations can be a lightning rod for Porsche aficionados. In a concours world where having the proper handle on a screwdriver in your toolkit is as important as world peace, one can imagine the disdain that some of these cars elicit. However, the Porsche 911 world has always been about self-expression and the cars were designed to be easily modified for track or street. It's only natural that some extremes would evolve in the marketplace.

SPECIAL EDITIONS & TUNER CARS
UTLIMATE PERFORMANCE: GT2 & GT3

No discussion of the modern-era Porsche 911 would be complete without including the Porsche 911 GT3 and GT2. Around 1998, Porsche retired its air-cooled platform with the 993, and released the 996. Based on the platform designed for the Boxster, the 996 was an improvement over the previous 911 in almost all areas, except one—the engine. When the Porsche 996 was introduced in 1999, Porsche took the cars to the track, but with terrible results. The "new" M96 engine installed in the street cars proved to be extremely unreliable on the track, and nearly all the cars experienced mechanical issues.

Then came the Porsche 911 GT3 and the GT2, Porsche's racing versions of the 996, available for the street and the track. The 996 GT3, first produced in 1999, was based on a chassis nearly identical to the street 996 cars, but with an older-style and more reliable engine. The GT3 engine was based on the proven engine of its predecessors, the Porsche 962 and Porsche 959. With a dry-sump system and a long-block (case, crank, rods) nearly identical to that of the 962/959, the GT3 was an instant success, particularly with customers who wanted to race a production car with minimal prep work. Porsche offered the car in a street version and a race-ready version that could be run in events like the Porsche Carrera Cup, GT3 Cup Challenge series, Porsche Supercup series, and FIA Formula 1 World Championship.

Similar in concept to the GT3, the GT2 is based on the Porsche 996 Turbo. The GT2 is best described as a two-wheel-drive racing version of the 996 Turbo, built in street and race versions as well. With about 475 horsepower combined with improved aerodynamics, the GT2 sits at the top of the Porsche 911 line-up as the ultimate 911, as current versions hold numerous track records.

ADDITIONAL READING: *Porsche 911 GT2 and GT3*, by Colin Pitt

SPECIAL EDITIONS & TUNER CARS
DAYTONA TRIBUTE: BRUMOS 997 & GT3

Peter Gregg founded Brumos Racing in 1971 to participate in the IMSA GT Championship series. Sponsored by Brumos Porsche, a Porsche dealership, the team achieved unparalleled success in American racing, winning the 24 Hours of Daytona four times. In the Porsche world, the Brumos Porsche livery—all-white cars, distinctive blue and red stripes, and usually the number 59—is synonymous with victory.

In honor of the legendary Brumos history, Porsche developed two commemorative models. The first is a Grand-Am-spec GT3 race car. In 2012, Porsche produced five of them with a 4.0-liter engine borrowed from the GT3 RS. The cars also included other features commonly found on Brumos's own race cars including a Motec engine management system, steering and wheel sensors, and a throttle-blipping system to assist with the sequential transmission. Decorated in the traditional Brumos livery, this GT3 is the ultimate ready-to-race track toy.

Also paying homage to the Brumos livery is the 2014 Porsche Carrera GTS B59 Edition. Specifically designed to salute Daytona champion Hurley Haywood (who won the 24 Hours of Le Mans three times, the 24 Hours of Daytona five times, and the 12 Hours of Sebring twice, all in Porsches), the B59 edition is a road-legal model based upon the 997. Adorned with classic 19-inch Fuchs-style alloys, blue and red interior trim, full sports suspension, a limited-slip differential, and Porsche Carbon Ceramic Brakes (PCCB), the B59 is one of the last and best models of the Porsche 997 line.

NORBERT SINGER

Nearly every historical photo of Norbert Singer shows him smiling. What's not to smile about when you're the brains behind of the most successful race car program of all time—the Porsche 956 and 962. In engineering circles, Singer is the hero behind Porsche's ultra-successful racing program. I even have a Norbert Singer action figure—dressed in Rothmans livery!

Singer arrived at Porsche in 1970 with an aerospace engineering background, useful skills that would prove invaluable in maneuvering around racing rules. Most notable was his rules-sidestepping design of the rear of the Porsche 935. The rules specified that the race car's rear window had to remain in place, which disrupted airflow over the car. Singer devised a clever workaround, having Porsche create a second window that fit on top of the original one. The results produced better airflow across the 935 silhouette and won Porsche five World Makes Championship titles.

Singer was also instrumental in designing the ground effects aerodynamics deployed on the 1982 956. The 956 was a breakthrough car that dominated motorsport racing for more than a decade. Its secret weapon? A newly designed airfoil that created a tremendous amount of downforce, which gave the car incredible cornering ability. The theoretical downforce from the 956/962 is so large at speed that it's greater than the weight of the entire car. In layman's terms, this means that the car could theoretically drive upside down in a tunnel at speed— the downforce from the ground effects aerodynamics would be strong enough to keep it stuck to the ceiling! While no one has actually attempted this, a 956 Rothmans car hangs upside down in the factory museum to commemorate this feat of engineering. Combining these revolutionary aerodynamic innovations with Hans Mezger's incredibly reliable flat-six engine led to an unstoppable racing program that won countless titles and ten world championships.

935: Porsche's legendary 911-based race car that dominated European and American racing in the mid-to-late 1970s.

Denloc: A new type of run-flat tire developed by Dunlop for the 959.

Ducktail: The distinctive rear spoiler that fits on top of the rear engine lid on the 1973 Porsche 911 RS.

Duesenberg: An American manufacturer of race cars and luxury automobiles in the 1920s. Duesenbergs were nearly all custom ordered and manufactured according to a client's wishes.

Fiberglass: A lightweight, glass-based material that is often used as bodywork for race cars.

Hollow-spoke: Wheels that are hollow-spoke have openings internal to each spoke. This allows for the design of a lighter weight wheel.

Kremer Racing: The successful Le Mans–winning team, who helped to popularize the conversion of street cars with slant nose conversions that resembled the team's Porsche 935 race cars.

Liter: A unit of displacement for engines. One to two-liter engines are considered small, while four-liter and greater engines are considered large.

Motegi: Twin Ring Motegi is a motorsport racetrack located in Motegi, Japan.

Nikasil: A protective coating used on the inside of cylinders on the higher performance Porsche engines.

Sheet metal: The sheets of metal (typically steel) welded together to form the chassis of a unibody car.

Speedster: Porsche's inexpensive model of 356 from the 1950s. Porsche released later versions of the Speedster based upon the 911 platform.

Supercup Series: An international motorsports racing series sponsored by Porsche that features identical Porsche 911 GT3 Cup cars.

Williams/Williams Grand Prix Engineering: A British Formula 1 racing car design firm and also an F1 racing team.

PORSCHE 911
COMMUNITY
AND CULTURE

PORSCHE 911 COMMUNITY AND CULTURE
STUTTGART

The Porsche crest, famous the world over, was first used on a Porsche in late 1952 and neatly integrates Stuttgart, Porsche's home, as part of its design. US Porsche importer Max Hoffman suggested a crest to Ferry Porsche, who drew the initial sketches. It was later finalized by Franz Xaver Reimspiess, who combined the traditional coat of arms of the former German state of Württemberg-Hohenzollern with the Stuttgart city seal. The crest has undergone only a few adjustments over the decades (mostly to the typefaces and bar colors) since it first appeared in the center of the 356 steering wheel.

Stuttgart was home to the Porsche "museum" that opened in 1976 and featured an exhibit space that could display about twenty cars at a time. Cars were rotated in and out of the museum, pulling from the 300 or so "reserve" vehicles stored offsite. While the quality of the cars on display was unquestionable, visitors frequently criticized the museum itself.

Stuttgart is also home to the Mercedes-Benz Daimler factory and museum. The Mercedes museum was completed in 2006 and it put Porsche's "hallway" museum to shame. Inspired by Frank Lloyd Wright's Guggenheim Museum in New York City, the Mercedes museum features a "double helix" design that maximizes space for the exhibits rimming the building.

Porsche responded to the Mercedes masterpiece by spending more than 100 million Euros to create a new facility that opened in 2009 and easily ranks as one of the world's premiere museums. Porsche had it relatively easy, though, only needing to create an impressive building, as the cars in the museum speak for themselves. With examples from Porsche's long and glorious racing history, and its beloved production cars, the museum ranks high on many auto enthusiasts' bucket lists.

PORSCHE 911 COMMUNITY AND CULTURE
R GRUPPE

Born from the roots of SoCal car culture, the R Gruppe is an invitation-only club of members focused on hot-rodding early 911s (1963–1973). It is named after the Porsche designation for some of its race cars (like the 911 R for Rennsport, or "racing" in German).

R Gruppe history stretches back to an article that appeared in the Porsche magazine, *Excellence*, in April 1998. The article profiled the early 1969 911 S modified by Cris Huergas, and shortly thereafter, fellow early 911 aficionado and famed car designer Freeman Thomas contacted Huergas. The two lamented the lack of opportunities for owners to modify their cars, and the idea for the R Gruppe was born. The club's premise is simple. As Freeman Thomas describes it, the club concentrates on three criteria. First, the cars must be built for "sports purposes," meaning they can be driven to the track, run for the day, and then driven home. Second, the cars are influenced by the SoCal hot-rod culture. And third, there's the Steve McQueen factor—a sense of style, restraint, being discreet and "cool."

The club has had a checkered reputation. Membership was originally limited to 300 individuals, leaving some on the outside feeling excluded. The concours-centric crowd that thinks the cars should be kept bone-stock sometimes winces at the modifications made to R Gruppe 911s. However, R Gruppe owners have inspired numerous creations that are now called an "R Gruppe 911," having achieved almost pop-culture status. Finally, there's a bit of a bad boy "I don't care what others think" attitude accentuated by loud pipes, fast cars, and long rallies through the canyon roads of Southern California.

PORSCHE 911 COMMUNITY AND CULTURE
PORSCHE CLUB OF AMERICA/CLUB RACING

One great aspect of 911 ownership is the supportive community surrounding the cars. The Porsche Club of America (the PCA) is a pinnacle point for that support with more than 100,000 members across twelve regions.

Local Porsche clubs (from both the Porsche Club of America and the Porsche Owner's Club) have extensive racing programs that encourage driver participation from beginner to advanced levels. All that is needed to race is a safety inspection, a helmet, a long-sleeved shirt, and closed-toe shoes. Of course, as one improves and moves up in the ranks, a dedicated track car becomes a necessary item to be placed on the holiday wish list.

One popular local club activity is autocross racing, timed racing trials typically held in a large parking lot, with a racecourse outlined by orange cones. Autocross is a fast-paced, exciting form of racing that's very easy to enter—most people can commute to the track and race their daily driver. It's a hands-on introduction to racing, and a good way to get familiar with the handling of a Porsche 911 on the track.

On the opposite end of the spectrum are Concours d'Elegance events. These competitive car shows let owners show off their cars and their restoration efforts. The goal is to show off the car in its best original condition within different classes. These range from the fun "Display Only" class, to the intense "Full Preparation" class where the judges evaluate every aspect of the car—trunk, chassis, engine bay, etc. This is what I call the Q-tip class, so named because all the participants are usually on the lawn the morning of the competition, cleaning their engine bays with Q-tips!

PORSCHE 911 COMMUNITY AND CULTURE
DUTCH NATIONAL POLICE PROWLER

FUN FACT

In 2014, Porsche lent the Australian police in New South Wales a new 911 Carrera that was dressed in police livery to attract younger recruits.

HISTORICAL TIDBIT

Most of these cars had standard "phone dial" or "cookie cutter" wheels, not the Fuchs alloys, presumably because the Fuchs alloys were an upgrade from the stock wheels, and not needed on a "government car."

KEY PERSON

California Porsche dealer Bill Yates saw the 1968 Porsche Polizei-Bekleidung 911 show car on display at American car shows in 1967. The car was a soft-window Targa and Porsche was convinced it would appeal to US police departments. Alas, Porsche did not achieve much success with police departments in the States, and Yates purchased the car and displayed it in his dealership showroom for several decades.

Perhaps the coolest Porsche "club" in the world was the Rijkspolitie, the Netherlands police division from 1945 through 1994. In the mid-1960s, Dutch police were tasked with finding a car fast enough to patrol the country's no-speed-limit highways. The cars had to be reliable, handle well, and have very good braking, and also needed an open top for greater visibility so officers could direct traffic while standing in the car. In the mid-1960s, only one car met these criteria, the Porsche 356.

The replacement of the 356 with the 911 caused some issues for the Rijkspolitie, as there was no open-top Porsche 911 available at its introduction. However, 1967 brought about the introduction of the 911 Targa, and the Dutch police embraced it. The 911 Targa became a department staple through 1993, when the switch was made to boring Volvos. In total, 507 Porsche cars wore the famous orange and white livery, easily establishing the Rijkspolitie as the largest historic Porsche police fleet.

The cars were typical production cars outfitted with special police equipment such as a blue flashing light on the Targa bar, a wooden box in the rear seat compartment to hold "police stuff" such as handcuffs and a breathalyzer, an illuminated "STOP" sign on the rear, extra front and rear lighting, and of course, a loud siren. All of the cars were painted white with bright orange trim and paint across the doors, hood, and rear decklid.

When they were retired from the force, most of the cars were stripped of the police equipment and sold as used sports cars in civilian spec. To restore one to its former squad car configuration involves tracking down old photographs, and gathering historic police equipment from the same era, not exactly an easy task.

PORSCHE 911 COMMUNITY AND CULTURE
RENNSPORT REUNION

FUN FACT

With more than 1,300 Porsches parked in the "Porsche Corral," a Rennsport Reunion is the perfect place to check out all the upgrades and modifications Porsche owners have made to their cars.

HISTORICAL TIDBIT

For 2018, the event was expanded to a four-day affair at Laguna Seca in central California. Porsche recently announced its intention to host the event every three years, with the next one planned for 2021.

KEY PERSON

Brian Redman, legendary Porsche race car driver, worked with Porsche Cars North America to organize the first Rennsport Reunion at Lime Rock Park in 2001.

Porsche 911 owners are among the luckiest car owners in the world. Not only do they enjoy the full support of Porsche's parts department a full sixty years after the 911 was introduced, but they also share in Porsche's heritage and racing history. Nowhere is this more evident than at Porsche Rennsport Reunion events held in the United States every three years or so.

Imagine a place where all of the race cars that you've seen and read about suddenly appear in the pits. Imagine them being driven by the same drivers that piloted the cars to victory at Le Mans, Sebring, and Daytona. Imagine seeing fifteen Porsche 962s battle it out on the track at Daytona or Laguna Seca—just as they did in 1987. If you're a true Porsche enthusiast, the Rennsport Reunion is not to be missed.

The first Rennsport Reunion event, held at Lime Rock Park in Connecticut in 2001, was sparsely attended as few people even knew about it. The next two events were held at Daytona International Speedway in 2004 and 2007, and made the reunion an instant bucket-list event for all Porsche owners. In 2011 and 2015, Rennsport Reunion moved to Laguna Seca in California—the land where Porsches are plentiful—and drew more than 40,000 attendees.

Pilgrimages to Rennsport Reunions in Porsche caravans are common, with local clubs organizing meets, rallies, and events leading up to the reunions. The factory has lent its full support to the events, showcasing the latest cars, hosting owner's events, offering test drives, and sponsoring banquets and dinners. Perhaps the most amazing part of each event is the concours, where all the race cars are displayed on the tarmac for all to view up close. No museum-style ropes or stanchions—the cars, owners, and drivers are inches away, available for incredible photos and autographs.

PORSCHE 911 COMMUNITY AND CULTURE
TRACK DAYS

"Win on Sunday, sell on Monday" is an old auto-industry phrase emphasizing the importance of auto racing to fuel new car sales. Porsche has a rich and successful racing history, built largely on the 911 platform. Porsche 911 owners know that when they buy a car, they will be able to drive it to the track and race it competitively with few modifications needed.

Certain elements of the 911 cement its reputation as a track-worthy car. The 911 suspension strikes a good balance between street and track performance and Porsche 911 braking performance has almost always been considered more than adequate for track days. But perhaps most important of all, the 911 engine's dry-sump oiling system, standard on all air-cooled 911s, was the same system Porsche used in its ultra-reliable race cars.

The best way to gain racing-level experience is through local clubs' track days. These events are often tailored for beginners, and also include driving instructors to help guide newcomers through the first track experience.

One of the more unique experiences available to Porsche owners (or anyone else for that matter) are the Porsche Experience Centers. These are private tracks that are located around the world in Atlanta, Los Angeles, Leipzig, Silverstone, Shanghai, Hockenheim, and Le Mans. Customers and fans can experience driving a Porsche with the assistance of a professional driving coach. The private courses feature small tracks with a variety of training modules ranging from skid pads to off-road courses, a quick-launch acceleration straight, and snow simulators. The Los Angeles Porsche Experience Center is also home to Porsche Motorsport North America, where private customers of Porsche can send their Porsche 962 or GT1 for service. It's not uncommon to see a handful of 962s or 959s in various states of repair at the center.

PORSCHE 911 COMMUNITY AND CULTURE
RESOURCES

The Porsche 911 is legendary in its own right, but what most owners don't realize is how extensive the amount of aftermarket, club, and factory support it continues to receive.

If you own an older car (such as a Ferrari 308, or an older BMW), you probably know parts for those cars can be difficult to obtain. The opposite is true of the Porsche 911. There is such a unique and passionate following for these cars that there remains a tremendous support community ready and waiting to supply knowledge, parts, and know-how.

Companies like Pelican Parts stock nearly every part needed for your 911. Stoddard NLA is another resourceful company that strives to recreate parts that have been "No Longer Available" (NLA) for many years. Even the factory, Porsche AG, maintains a very large supply of parts for even the oldest 911s, and it recently launched a new "Porsche Classic" program to reproduce more NLA parts for older cars. Very few manufacturers offer this level of support for their cars—it enables the enthusiast to drive the heck out of the car and not to worry about breaking precious parts that are "unobtanium."

The support network for these cars is nothing short of amazing. Websites such as the Pelican Parts Forums and Rennlist offer millions of posts and discussions on nearly every aspect of modifying, repairing, and restoring older 911s. More history and how-to books have been written about the Porsche 911 than any other car, including three best-sellers by your faithful author (such as *101 Projects for Your Porsche 911*). Finally, many magazines support the marque and the 911 in particular, including *Excellence Magazine, 000, Total 911*, and *911 and Porsche World,* to name a few.

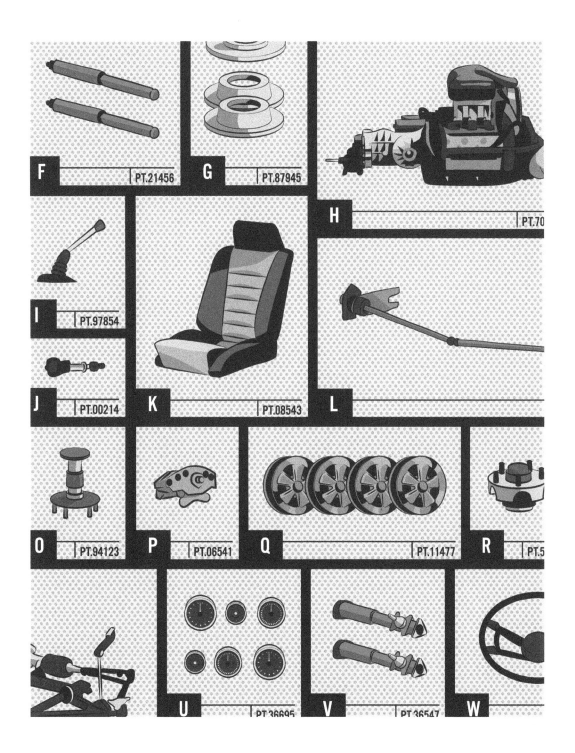

F PT.21456

G PT.87945

H PT.70

I PT.97854

J PT.00214

K PT.08543

L

O PT.94123

P PT.06541

Q PT.11477

R PT.5

U PT.36695

V PT.36547

W

To automotive performance aficionados, there are few cooler plac-es on Earth than the Nürburgring racetrack in western Germany. Nicknamed "the Ring," Nürburgring consists of a few different loops that are sometimes open to the public. The North Loop, or Nordschleife, is considered to be the main ring and is very popular with "sporting tourists," or Touristenfahrten, who like to test their driving skills in their performance cars. This loop runs amid mystical green Eifel forests that are often shrouded in mist that creates wet road conditions. It is affectionately known as the "Green Hell," a nickname coined by racing great Jackie Stewart.

While there is technically no speed limit on most of the ring, the entrance ticket officially states that racing and speed record at-tempts are forbidden—guidelines that drivers commonly ignore. Driving a sports car (ideally a Porsche 911) around the Nürburgring Nordschleife is a dream of many automotive enthusiasts. One would only hope that they would not become a member of the infamous "Bongard Club"—so named after the company that operates the towing service that quick-ly removes crashed cars from the ring.

Porsche made history at the ring on May 28, 1983, when a twenty-five-year old Porsche driver named Stefan Bellof piloted a Porsche 956 around the Nürburgring Nordschleife in 6:11.13—a record that will likely never be beaten. Why? Because the following year, the track length was shortened from 14.2 miles to 12.9 miles, forever changing the layout and solidifying the 956 as the fastest car ever around the ring.

Porsche officials often regard Nürburgring as their backyard test track. Most recently in 2017, Porsche set loose a 911 GT2 RS that broke the track record for the fastest time in a production car (even beating Porsche's previous record set in a 918 Spyder). Advances in tire tech-nology, combined with advanced aerodynamics that generated 750 pounds of downforce, helped produce a top speed of 211 miles per hour. Running the spoilers at the maximum angle of attack to produce 992 pounds of downforce allowed the GT2 RS to hit 193 miles per hour on the Nürburgring's final straight. Technology will continue to advance, but for now, a Porsche 911 holds the track speed records.

GLOSSARY: PORSCHE 911 COMMUNITY AND CULTURE

Autocross: A timed race, typically staged in a large parking lot, where drivers have to maneuver around a tight course, often one defined by traffic cones.

Bongard Club: The nickname given to drivers who crash their car at the Nürburgring.

Concours d'Elegance: A contest where cars are judged and honored for being in the best possible original condition.

GT1: Porsche's successor to the 962 race car.

Laguna Seca: A racetrack located in central California near the Pacific Ocean.

Lime Rock Park: A racetrack located in Salisbury, Connecticut.

Mercedes-Benz Daimler: An automotive manufacturer located in Germany, also headquartered in Stuttgart.

NLA: "No Longer Available," typically meaning the manufacturer does not intend to ever produce the part again.

Nordschleife: The north loop of the Nürburgring racetrack.

R Gruppe: A 911-centric owner's club that concentrates on hot-rodding the 1965–1973 911.

Recaro: A German manufacturer of performance seats.

Soft-window Targa: The 1967 Porsche 911 Targa model that featured a fabric and plastic rear window.

Steve McQueen: A movie actor from the 1950s to 1980 who owned many Porsches and was an unofficial spokesman for the brand. McQueen starred in and produced the movie *Le Mans*, which featured the Porsche 917.

Stuttgart: The capital of the German state of Baden-Württemberg and home to Porsche's corporate headquarters.

INDEX

LOOK FOR OTHER TITLES IN THE
SPEED READ SERIES

SPEED READ F1
Stuart Codling
ISBN: 9780760355626

SPEED READ CAR DESIGN
Tony Lewin
ISBN: 9780760358108

SPEED READ FERRARI
Preston Lerner
ISBN: 9780760360408

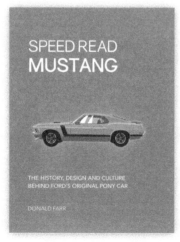

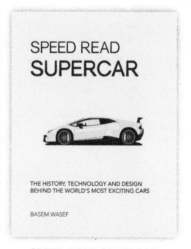

SPEED READ MUSTANG
Donald Farr
ISBN: 9780760360422

SPEED READ SUPERCAR
Basem Wasef
ISBN: 9780760362914

Milton Keynes UK
Ingram Content Group UK Ltd.
UKHW020655150624
444068UK00006B/46